# *Vegetables*

# VEGETABLES

**A BIOGRAPHY** | *Evelyne Bloch-Dano*

TRANSLATED BY TERESA LAVENDER FAGAN

*The
University
of Chicago Press
Chicago and
London*

EVELYNE BLOCH-DANO is the author of many books, including a prize-winning biography of Émile Zola's mother, as well as *Madame Proust: A Biography* and most recently a novel, *Le Biographe*.

TERESA LAVENDER FAGAN has translated over a dozen books for the University of Chicago Press, including J. M. G. Le Clézio's *The Mexican Dream* and most recently Mustafa Chérif's *Islam and the West: A Conversation with Jacques Derrida*.

The University of Chicago Press, Chicago 60637
The University of Chicago Press, Ltd., London
© 2012 by The University of Chicago
All rights reserved. Published 2012.
Printed in the United States of America

20  19  18  17  16  15  14  13  12      1  2  3  4  5

ISBN-13: 978-0-226-05994-5 (cloth)
ISBN-10: 0-226-05994-4 (cloth)

Originally published in French as *La Fabuleuse histoire des legumes*
© Editions Grasset & Fasquelle, 2008.

Cet ouvrage a bénéficié du soutien des Programmes d'aide à la publication de Culturesfrance/Ministère français des affaires étrangères et européennes. * This work, published as part of a program of aid for publication, received support from CulturesFrance and the French Ministry of Foreign Affairs.

Ouvrage publié avec le soutien du Centre national du livre–ministère française chargé de la culture. * This work is published with support from the National Center of the Book–French Ministry of Culture.

Library of Congress Cataloging-in-Publication Data

Bloch-Dano, Evelyne, author.
[Fabuleuse histoire des legumes. English]
Vegetables : a biography / Evelyne Bloch-Dano; translated by
Teresa Lavender Fagan.
    p.  cm.
Includes bibliographical references.
ISBN-13: 978-0-226-05994-5 (cloth: alkaline paper)
ISBN-10: 0-226-05994-4 (cloth: alkaline paper) 1. Vegetables—History.
I. Fagan, Teresa Lavender, translator. II. Title.
SB320.5.B6613 2011
635—dc22                              2011004425

♾ This paper meets the requirements of ANSI/NISO Z39.48-1992 (Permanence of Paper).

*vegetable* The early use was adjectival in the sense "growing as a plant," from late Latin *vegetabilis* "animating." The noun dates from the late 16th century. Related words include *vegetative*; *vegetation*; and *vegetate*. The slang use *veg out* meaning "pass the time in mindless activity" arose in the 1980s. *Vegetarian* is an irregular formation of the mid-19th century; the abbreviation *veggie* dates from the 1970s. :: From *The Insect That Stole Butter? The Oxford Dictionary of Word Origins*, 2nd ed., edited by Julia Cresswell (Oxford University Press, 2009)

# Contents

## TRANSLATOR'S NOTE

As the reader will learn from Michel Onfray's remarks (see "A Biographer of Vegetables" at the end of this book), Evelyne Bloch-Dano's text comes out of presentations she made at the Université populaire du goût, addressed to a predominately French audience. Thus much of the text includes references to French authors, proverbs, customs, and so forth, which may not be easily accessible or understandable to an Anglophone reader. I have added footnotes wherever I felt the text needed explanation. And with the author's input and approval, I have replaced some of her material with more Anglo-friendly alternatives. For excerpts from better-known French texts, where possible I've provided published English-language translations with citations; otherwise, I've provided my own translations. For song lyrics and poetry, I've maintained the original French and provided my literal translations in footnotes.

I wish the reader *bonne lecture* and *bon appétit!*

*"In the end, it's all about love."*
JOSEPH DELTEIL

# *Introduction*

✤ ✤ ✤ ✤ ✤ ✤ ✤ ✤ ✤ ✤ ✤ ✤ ✤ ✤ ✤ ✤ ✤ ✤ ✤ ✤ ✤ ✤ ✤ ✤ ✤ ✤

We used to spend the end of the summer at my grandparents' home. The scrawny little Parisian girl—they nicknamed me "the string bean"—was finally going to put on some weight thanks to my grandmother Rosa's *tarte aux mirabelles* and *pot-au-feu!\** But gobs of fat in the soup made me sick, and I didn't like desserts. Only the raw potato slices that she roasted on the stove and the sautéed veal sweetbreads, the likes of which I have never, ever encountered, could tempt my taste buds. Grandma watched me sternly and sadly as I picked at my food. No sooner had she turned her back than I ran off to find Bouboule, the dog that was the love of my childhood life.

My grandparents owned a wholesale candy business. They used their garage to store the merchandise: mountains of boxes filled with Suchard chocolate bars; pink, blue, or white *dragées*; lemon drops; round tins of yellow cocoa, which was dissolved in water; strips of black licorice; jam-filled candies; Mistrals, both winning and losing;† vanilla wafers. These treasures didn't tempt me.

But at the back of the courtyard was a shed where they stored empty wooden crates, empty boxes, bicycles, rusty tools, all the dregs of the household. The ground was damp, and it was dark and cool in there. It was Bouboule's and my retreat. Scarcely taller than the old mutt, I

---

*\*The former, a tart made with sweet yellow plums; the latter, a beef stew.*

*†A Mistral Gagnant is a candy. Some are "winning" (gagnant), meaning that you can get another one for free.*

played among the crates, and when I was hungry or thirsty, I shared the leftovers soaking in a mixture of greasy water and milk in his enamel dish. The triumphal beginnings of a gastronome . . .

The shed opened onto an enclosed little garden with thick walls. When the weather was good, I went and hid there with Bouboule.

Like other inhabitants of that Lorraine town, my grandparents had a real vegetable garden outside the village. We would walk past the farm where we went to get milk, and then take a path bordered by blackberry bushes and little plum trees. It was there, years later, that I fought the boys who ambushed girls. I loved those fights; I came back covered with bruises, my knees bloodied, flushed with excitement. The gardens were laid out on either side of the path. They were tended by Madame Lipp; Grandma plied her at noon with huge plates of Alsatian sauerkraut and in the afternoon with café au lait and *tartines* (hunks of bread and jam).*

I don't think anyone ever went into the little garden at the back of the shed. It was full of weeds, along with one clump of rhubarb and a thick row of curly parsley. I remember moments of pure ecstasy there, sitting on the ground, my back against the wall warmed by the sun. I would pull up parsley by the handful and munch it. My mouth was filled with the tingly flavor; I chewed the leaves and even the rather stringy stems, and the juice dribbled down my chin as I faded into delightful intoxication. Bouboule, lying next to me, perked up his ears from time to time: someone was calling me from the courtyard. I heard, yet I didn't hear. I was in the abandoned vegetable garden, and I felt wonderful. Out of reach, alone, happy in the parsley.

There have been other gardens in my life: on vacation rental properties, at houses I've stayed in; the garden of the "little lady" in Seine-et-Marne who sold us lettuce and green beans; the garden of a grandmother from Poitou I once knew, its paths bordered with carnations and thyme. My father planted flowers everywhere, whether the house belonged to him or not. In the spring we would go to pick wild hyacinths in the Saint-Germain-en-Laye forest, lilies of the valley in Chaville, and lilacs in the Chevreuse valley. Then my parents bought a small house with a garden in the Oise département, and its flower beds were admired by

---

*Often oven-toasted and buttered or spread with jam, *tartines* are typically eaten in France for breakfast or as an afternoon snack.

all who saw them. A white lilac bush with double blooms bore witness to their new status as property owners. But there were no vegetables, apart from a few herbs. Flowers, yes, in abundance. I can still see my father on his knees, wearing the blue overalls reserved for that purpose, trowel in hand. As he grew older, it was harder for him to stand up again. My mother, who wouldn't have put her beautiful hands in the dirt for anything in the world, assumed the task of assembling bouquets from the flowers wrapped in damp newspaper that we brought back to Paris on Sunday night.

There are only two or three rosebushes and a few bellflower plants left from the time when the house was sold. The new owners pulled up the white lilacs to build a porch.

My father was buried without flowers or wreaths.

Growing up in a small fourth-floor apartment in Paris hardly fosters a green thumb. As for a vegetable garden, all I could do for quite a while was to plant lentils and beans in damp wads of cotton. Wonder of wonders, when I was ten, standing in front of the whitish stalk and the first tiny, translucent green leaf, I saw it was growing! Alas, inevitably, the plant shriveled up and withered as soon as it was put into a pot on the windowsill. It ended up dying. I never understood why.

And so my gardens are rooted in the realm of the imagination. They sprouted in the writings of three women who, each in her own way, loved the countryside and inspired me to love it too: the Countess of Ségur, Colette, and George Sand. Places respectively devoted to games, to motherhood, and to creation: the small garden of perfect little girls, the vegetable garden of Sido, and the formal gardens of the lady of Nohant served as models, at various ages, for my country dreams—and sometime for my utopias.

I had to wait for my first house in Normandy, which I shared with my companion at the time, to finally have a flower and vegetable garden. My father—again—had warned me: "The people from around here will respect you only if you take care of it!" I made an effort, but . . . The rows were uneven or curved, the plantings irregular; there were years when nothing grew, years of wormy carrots; I had no style, routine, or technique. I've improved a bit, but I still have a lot to learn. I lack the primary quality of a gardener: patience. I like to sow, plant, and harvest, but when it comes to cultivating. . . . I lived for more than ten years in that unheated old farmhouse. Winters are long and harsh in the Orne. From those years I developed a tendency to hibernate and an obsession with warmth. Later, with Pierre, there was a small vegetable garden on the balcony of our apartment in Bougival (we had strawberries and lettuce), and we tried to start one in our garden in L'Étang-la-Ville, where only roses, cherries, and raspberries really seemed happy.

Now, in our vegetable garden in the Orne, we grow strawberries, peas, squash, leeks, lettuce, cabbage, beets, cucumbers, artichokes, tomatoes, garlic, onions, and potatoes. Sage, rhubarb, sorrel, thyme, savory, tarragon, and chives feel at home there. Basil, always cautious, stays in its pot. For here I am again, years later, in Normandy, connected to it by the very roots of my heart. This time our country house is heated.

Without Normandy, then, I wouldn't have known about picking beans in the summer heat, or lettuce powdered with dirt, or green tomatoes, or giant, hollow squash, or cabbage with leaves full of holes. Nor about red beets that can be bitten into like a delicacy, peas that melt in your mouth, juicy onions, green beans al dente, strawberries warmed by the sun, meaty tomatoes, the curve of a melon. Without Normandy, I wouldn't have known the happiness of tasting my own vegetables, the most beautiful and delicious in the world, of course!

And without Michel Onfray, my friend and longtime cohort, I would never have become involved in the adventure of the Université populaire du goût (UPG).* I would never have woken up at night wondering how I was going to do this. How could I go from Proust to carrots, from biographies of women to biographies of vegetables? I would never have concocted these jardinières, mixing the history of the vegetable with that of taste, literature with botany, art history with the history of food. I would never have been able to study that vast intermingling of cultures, from the sources of civilization to the most recent findings in genetics, from people of the Neolithic to children of junk food, from China to the Andean plateaus, from the Middle East to Canada, from the delta of the Indus to the arid regions of Africa. I would never have had the unique experience of talking about tomatoes or peas to an interested audience of several hundred. I would never have experienced

*For more on the "Popular University of Taste," see Michel Onfray's "A Biographer of Vegetables," below.

this wave of friendship, the camaraderie, the "*Ohs!*" and "*Ahs!*" at the marvels of the chefs. I would never have met the people who work in the Jardins dans la Ville, the "Gardens in the City," an association that reintegrates the marginalized through urban garden projects, or joined them in passing around platters of foie gras with Jerusalem artichokes. And then, let's admit it, without the UPG, I would have known nothing about cardoons. I might have thought a parsnip was a type of turnip.

And yet, when I really think about it, from my grandmother's vegetable garden to ancient vegetables, from the biographies of women to the beginnings of the tomato, from dreamed-of gardens to the Norman countryside, from writers' houses to histories of taste, the path was obvious. It passed through my curiosity, my love of food, my love of literature, nature, and life. And it was enriched by my experience. Peeling and grating carrots; cutting up leeks and turnips for soup; stuffing tomatoes, squash, eggplants, or green peppers; shelling peas or beans; washing lettuce; mashing potatoes—all of these are everyday activities, simple and noble. They were passed on to me, and I am happy to pass them on to my children.

With the UPG, I rediscovered this pleasure of passing things on. The following "histories" should not be considered either erudite or exhaustive. They represent my desire to offer my listeners, and now my readers, some stories of vegetables that I have harvested here and there— and have prepared according to my own recipes.*

---

*AUTHOR'S NOTE: *The texts of my lectures have been revised, expanded, and edited. The first sessions of the UPG took place from December 2006 to June 2007. Each lecture on a given vegetable was followed by demonstrations by the chefs. I have maintained the order of the presentations. The chapters on beans and squash are part of my 2007–8 course "Histoires de goûts," which includes contributions from plant and nutrition specialists. The chapter on the chili pepper is added just for spice!*

Lettuces, endives, chicory, open and with rich soil still clinging to their roots, exposed their swelling hearts; bunches of spinach, sorrel, and artichokes, piles of peas and beans, mounds of cos lettuces, tied up with straw, sounded every note in the scale of greens, from the lacquered green of the pods to the coarse green of the leaves; a continuous scale of rising and falling notes that died away in the mixed tones of the tufts of celery and the bundles of leeks. But the highest notes, at the very top of the scale, came from the bright carrots and snowy turnips, scattered in tremendous quantities throughout the markets, which they lit up with their medley of colors. At the intersection in the rue des Halles, mountains of cabbages were piled up; there were enormous white ones, as hard as cannon balls, curly ones with big leaves that made them look like bronze bowls, and red ones which the dawn seemed to transform into magnificent flowers with the hue of wine-dregs, splashed with crimson and dark purple. On the other side of the markets, at the intersection near Saint-Eustache, the opening to the Rue Rambuteau was blocked by a barricade of orange pumpkins in two rows, sprawling at their ease and swelling out their bellies.

Émile Zola, *The Belly of Paris*

TRANSLATED BY BRIAN NELSON, PP. 25–26

*A Little History*

GOOD TO

# *Eat,*

GOOD TO

# *Think About*

✤ ✤ ✤ ✤ ✤ ✤ ✤ ✤ ✤ ✤ ✤ ✤ ✤ ✤ ✤ ✤ ✤ ✤ ✤ ✤ ✤ ✤ ✤ ✤ ✤

Nourishment is central to living, from the air we breathe to the food we ingest; nourishment is what keeps us alive, but also what connects us to our environment, our history, our society, our times, our social status— to others. For Claude Lévi-Strauss, "the cooking of a society is a language into which it unconsciously translates its structure, or else resigns itself, still unconsciously, to revealing its contradictions": as a touchstone for our behavior and beliefs, our myths, our systems of organization, cooking is not only "good to eat," but also "good to think about."

To nourish oneself, to eat . . . "Tell me what you eat and I will tell you what you are," wrote the nineteenth-century gastronome Jean Anthelme Brillat-Savarin at the beginning of his *Physiologie du goût* (1825). But tell me what you eat, and I will also tell you what connection you have with your loved ones, with nature, with culture, with society. By feeding ourselves, we are not just addressing our corporeal envelope (thin, fat, too thin, too fat) but also our brain, our senses, our psyche. Humans are the only living beings not to mechanically submit to the constraints of their environment, but to be able to choose their food according to criteria that are not physiological but symbolic. They might favor this symbolic dimension to the detriment of their health or life: food seen as *mana* that gives life but can also cause death; foods that are taboo, totemic, whose substance is incorporated; rituals, forbidden or made sacred, from the young Hua of Papua New Guinea who eat a fast-growing vegetable to grow faster, to the Jew or the Muslim who refrains from eating pork, from the Indian Brahman who will not eat the flesh of a cow, to the Christian who symbolically absorbs the body of Christ through the host. But there is a symbolic dimension in all the great gatherings in life that are marked by rituals: marriage or anniversary meals, Saint Sylvester feasts or funeral lunches. Companions, friends, are those with whom one breaks bread.

✤ ✤ ✤ ✤ ✤ ✤ ✤ ✤ ✤ ✤ ✤ ✤ ✤ ✤ ✤ ✤ ✤ ✤ ✤ ✤ ✤ ✤ ✤ ✤ ✤ ✤

> And God said, Behold, I have given you every herb bearing seed, which is upon the face of all the earth, and every tree, in the which is the fruit of a tree yielding seed; to you it shall be for meat. ✳ GENESIS 1:29

✤ ✤ ✤ ✤ ✤ ✤ ✤ ✤ ✤ ✤ ✤ ✤ ✤ ✤ ✤ ✤ ✤ ✤ ✤ ✤ ✤ ✤ ✤ ✤ ✤ ✤

Today food is of interest to anthropologists, archaeologists, archaeo-zoologists (who teach us, for example, that contrary to what is believed, the Gauls ate less wild boar than domesticated beef or horse), historians (through the study of table manners or the history of taste and food), psychologists, linguists (here's a riddle: what French word has the same derivation as the word "pudding"?),* geographers, economists, sociol-ogists, botanists, doctors, biologists, and politicians. Terrain, climate, and natural resources long influenced the culinary destiny of a region or a people. But wars and migrations have also played a role in alimen-tary changes by introducing new vegetables that have become essen-tial elements in the regional diet, such as the potato, which came with the conquistadors from South America to Europe, or the soybean with the Chinese to Japan. And let's not forget that human history since its beginnings has been marked by food shortages and famine. It doesn't take long for a population to become malnourished through a change in climate or a war: European citizens haven't forgotten the last one—scarcity or even lack of butter, eggs, milk, sugar, meat, coffee, vegeta-bles. More recently, inhabitants of Darfur, Ethiopia, and Somalia have known starvation. Food has settled into the heart of politics through such issues as sustainable development, environment, exchanges be-tween North and South, or genetically modified crops.

    As for social "distinction," so dear to the sociologist Pierre Bourdieu,

---

*AUTHOR'S NOTE: Answer: le boudin, *a sausage of Viking origin.*

nowhere is it manifest with as much consistency as in matters of food, from table manners to food products, and to cooking itself. Globalization has not brought equality. Granted, we eat better, and China, for example, has gone from an endemic quasi-famine to a national diet that in thirty years has caused an average three-inch increase in the height of their children. But everywhere, and most notably in the developed countries, qualitative differences in kinds of food remain—and may even have been exacerbated. From the meal trays on long-distance flights to the "gourmet sections" in supermarkets, not everyone eats first-class. On the one hand, there is "fine cuisine," on the other, "chow." Claiming to be a gastronome is already a distinctive sign of belonging. What is gastronomy if not the fruit of a nineteenth-century collaboration between writers and food professionals to help the bourgeoisie attain a culinary art that until then had been reserved for the nobility? This doesn't mean that members of the middle or lower classes ate less well, but that they ate differently, and more important perhaps, that they *talked about it* differently. Simply reading restaurant menus is enough to be convinced of this.

———

And because vegetables connect us to the earth, to that Mother Earth of whom the ancients spoke, they occupy a very specific place in the history of food, as well as in our imaginations, our myths, our customs, our family heritages. They have long constituted if not the foundation of food, as assured by grain, at least the most elementary part. From picking to gathering—even today, dandelions in the fields, mushrooms in the woods, blackberries along woodland paths—vegetation has sure value, one that guarantees subsistence when one has nothing. Vegetables were at the dawn of humanity; they form the elementary degree of social organization, the passage from the raw to the cooked, from nature to culture, from the stage of gathering to that of cultivation. Humans have tamed vegetables the way they have domesticated animals, by selecting plants and observing the effects of those plants on their bodies. Plants, grains, herbs, and roots follow the beginnings of sedentarization: planning to grow something assumes that one is settling down for the time it takes to plant, allow plants to grow, and harvest the bounty.

The vegetable, like anything that is grown, is associated with time, patience, the rhythm of the seasons. Gardeners know this through experience—an experience that has become a symbol of human existence, as highlighted in Ecclesiastes 3:2: There is "a time to be born, and a time to die; a time to plant and a time to pluck up that which is planted."

There are spring, summer, autumn, and winter vegetables. Just as it takes nine months for a baby to come full term, it takes a certain amount of time for a cabbage to grow, or a tomato to ripen. Chemistry has modified natural development, and it is no coincidence that our civilization of speed, performance, and consumption aims to produce vegetables more rapidly, bigger, more perfect, better balanced—redder tomatoes and greener lettuce, endives whiter than nature intended. Culture is winning over nature.

Napoleon III was delighted that peas could arrive fresh in Paris thanks to the railroad; thus vegetables also speak to us of space. The space of the vegetable garden: patches marked with strings, impeccable geometry, alignment, order, judicious spatial economy. But also a social space, contrast or even rivalry between villages, or, on the other hand, the exchange of seeds, recipes, and produce, transmitted from generation to generation. When the chain is broken, what happens? Varieties and even species disappear. In wartime, the gap between rural and urban space lessens, yet becomes more manifest. Thus, during World War II, Jerusalem artichokes and rutabagas fed urban populations that no longer had the luxury of potatoes. Surreal vegetable gardens appeared during the bombardment of Berlin in 1945, when women planted a few vegetables in gutted apartments without doors or windows, to stave off starvation. The vegetable thus incarnates life's revenge over death, the triumph of freshness over decay, victory of the rural over the urban.

At the end of the arbor, near the plaster lady, stood a kind of log cabin. Pécuchet stored his tools in it, and there he spent delightful hours husking seeds, writing labels, arranging his little pots. To take a break, he sat on a crate by the door and planned improvements to the garden.

He had made two bushels of geraniums for the foot of the steps. Between the cypresses and the cordons he planted sunflowers. And as the flower beds were covered in buttercups, and all the alleyways in fresh sand, the garden was resplendent with a variety of yellow tones. But the hotbed was soon crawling with larvae; and despite the insulation of the dead leaves, beneath the painted frame and slathered lids the growth was sickly to behold. The cuttings didn't take; the grafts came undone; the sap in the layers stopped flowing; the trees had white spots on their roots; the seedlings were a desolation. The wind enjoyed flattening the beanstalks. The abundance of sludge ruined the strawberries, the lack of pinching killed the tomatoes.

There was no broccoli, eggplant, turnips, or watercress, which he had tried to grow in a tub. After the thaw, they lost the artichokes. The cabbages were his only consolation. One in particular gave him hope. It blossomed, grew, ended up being huge and absolutely inedible. No matter: Pécuchet was glad to have produced a monster. ✳ GUSTAVE FLAUBERT, *Bouvard and Pécuchet,* translated by Mark Polizzotti (Dalkey Archive Press, 2005), pp. 28–29

✤ ✤ ✤ ✤ ✤ ✤ ✤ ✤ ✤ ✤ ✤ ✤ ✤ ✤ ✤ ✤ ✤ ✤ ✤ ✤ ✤ ✤

But vegetables also point to the lot of the poor in the social history of food. In the seventeenth century, La Bruyère saw peasants as "wild animals" who "retire at night into their hovels, where they live on black bread, water, and roots." Grain, in whatever form, was the foundation and symbol of poverty, and meat was long a sign of wealth and luxury. Vegetables, if leafy and non-leguminous, were believed not to be nourishing. Unlike grains, they did not leave the body feeling full, close to satiety. Further, for a long time vegetables were gastronomy's poor cousins. They served a useful purpose as side dishes or garnish. They decorated meat, game, or fish, functioning merely as enhancement. For Grimod de La Reynière,* author of *L'Almanach des gourmands* (1803), "the man who is truly worthy of the title of gourmand scarcely views

---

*Alexandre Balthazar Laurent Grimod de La Reynière (1758–1837), who was trained as a lawyer, acquired fame during the reign of Napoleon for his sensual and public gastronomic lifestyle.*

vegetables and fruits except as means to clean his teeth and refresh his mouth, and not as products capable of feeding a hearty appetite." The vegetable is not a noble item; it is humble, ignoble in a literal sense. It is probably for that reason that it figures less in poetry and art (with the exception of still life) than do fruit, flowers, and trees. There is no lyric poetry for vegetables, except in the familiar space of fables, short odes, or burlesque diversions, such as Ronsard's poem dedicated to salad.* Metaphors confirm this low status: the French say *belle comme une rose,* "beautiful as a rose," and *bête comme chou,* "dumb as cabbage"; *un teint de lys,* "the complexion of a lily," or *une peau de pêche,* "skin like a peach," but *une mine d'endive ou de navet,* "the look of an endive or turnip." Being likened to a gazelle is one thing; to an asparagus, quite another. To call someone a vegetable evokes not the vegetal but the vegetative. As for an important man, he is called, derisively, *une grosse legume,* "a big vegetable."

✧ ✧ ✧ ✧ ✧ ✧ ✧ ✧ ✧ ✧ ✧ ✧ ✧ ✧ ✧ ✧ ✧ ✧ ✧ ✧ ✧ ✧ ✧ ✧

### RECIPE FOR A SALAD

To make this condiment, your poet begs
The pounded yellow of two hard-boiled eggs;
Two boiled potatoes, passed through kitchen sieve,
Smoothness and softness to the salad give.

Let onion atoms lurk within the bowl,
And, half suspected, animate the whole.
Of mordant mustard add a single spoon,
Distrust the condiment that bites so soon;
But deem it not, thou man of herbs, a fault,
To add a double quantity of salt.

Four times the spoon with oil from Lucca crown,
And twice with vinegar procured from town;
And, lastly, o'er the flavored compound toss
A magic soupçon of anchovy sauce.

---

*For the Ronsard poem that appears in the French edition, I have substituted one by the English writer Sydney Smith.

O green and glorious! O herbaceous treat!
'Twould tempt the dying anchorite to eat:
Back to the world he'd turn his fleeting soul,
And plunge his fingers in the salad bowl!
Serenely full, the epicure would say,
"Fate cannot harm me, I have dined to-day."

✳ SYDNEY SMITH (1771–1845)

✣ ✣ ✣ ✣ ✣ ✣ ✣ ✣ ✣ ✣ ✣ ✣ ✣ ✣ ✣ ✣ ✣ ✣ ✣ ✣ ✣ ✣ ✣ ✣ ✣

In modern times, the vegetable seems to have become a luxury in developed countries, considered a light choice in contrast to common, stodgy food. On the one hand, we have pastas and "spuds"; on the other, green beans, lettuce, and the like. In other words, noodles go with grub, and lettuce (a metaphor for money in French) with sorrel. On the one hand, there are those who feed; on the other, those who eat; or, to quote Brillat-Savarin again, "Animals feed, man eats; the thinking man alone knows how to eat." But everyone can know how to eat—because knowing how to eat can be learned.

So today, the fight against poor eating habits begins with vegetables.

In the collective consciousness, vegetables are associated with diets and austerity. For Rabelais or Rousseau, they were the key to a balanced and sober eating plan. We find them in the maternal or medical commands "Eat your spinach," "Eat your veggies if you want to grow tall," and in the recommendation to eat five servings of vegetables a day to limit the risk of cancer or cardiovascular disease. And so vegetables are considered to be utilitarian, not pleasurable, classified with dietetics and not with gastronomy, with rational adulthood and not childhood delights. They are contrasted with the regressive food that fills lazy stomachs. Obesity is a social marker: a sign of wealth in the past when many never had enough to eat; but in our Western societies, where slenderness is an aesthetic goal, obesity preys on the less fortunate. Too many people eat fatty foods, with an excess of salt or sugar, glucides or lipids—snack foods, prepared meats, chocolate bars, fast food, chips, TV dinners, pizza, panini, greasy hero sandwiches, mayonnaise, ketchup.

Children are overfed, stuffing themselves at any hour of the day following the principle of instant gratification, and thus growing fat. Many adolescents, too, fall onto that path of easy access, which is perhaps even more disturbing. Recent figures indicate that one out of four teenagers in France is threatened by obesity.

✣ ✣ ✣ ✣ ✣ ✣ ✣ ✣ ✣ ✣ ✣ ✣ ✣ ✣ ✣ ✣ ✣ ✣ ✣ ✣ ✣ ✣

Vegetables, fretful vegetables. I think of those women whose lives have slipped away in the shadow of your peels, and the knife shone, dancing, in their hands. Vegetables, your sad colors, your peels, your texture. I certainly would not mince the salsify of reproaches, the rutabaga of you-see-what-I-means, the turnip of endless excuses, the radish, the radish. I will leave the tomato-faced objections, the artichoke-heart quibbles, the squash squeals, the complete critical horseradish, to the molds. ✳ LOUIS ARAGON, *Treatise on Style*, translated by Alyson Waters, p. 107

✣ ✣ ✣ ✣ ✣ ✣ ✣ ✣ ✣ ✣ ✣ ✣ ✣ ✣ ✣ ✣ ✣ ✣ ✣ ✣ ✣ ✣

And vegetables demand that we take our time with them. They have to be bought, peeled, washed, cut up, cooked. They wilt, darken, soften, rot. To prepare them takes time, attention, creativity. But those are the very reasons why they can be a source of pleasure and not just a necessity.

For vegetables don't simply connect us to the land that produced them; they also grow on the enriched land of our affective memories: a mother's leek soup, an aunt's artichoke and bean tagine, a grandmother's stuffed cabbage, the lentils in the school cafeteria. Each one of us has our own evocative memories, those flavors that speak to our hearts. Vegetables belong to what is intimate, family cooking, home traditions. They are not fancy but comforting—carrots, turnips, and leeks simmering in the soup pot, jam bubbling in the preserving pan, an apple pie in the oven.

To speak of vegetables, then, is to travel in search of a territory, a culture; it is to rediscover the traces of a history that weaves in and out of the etymology of a word, the travels of a plant from one region to another, from one country to another, from one symbolic realm to another (why do carrots make you likable, and why are children born in the cabbage patch?);* from a vegetable garden to a poem; from a painting to a market woman pushing her cart though the streets of Paris while yelling the praises of her fresh lettuce; from a song to a conquistador bringing back seedlings and new condiments in the hold of his caravel. It is to travel through space and time, from the collective to what is most intimate; it is to intersect our knowledge and our questions, our experiences, our curiosity. In the shell of a pea, in the seeds of a tomato, in radish tops that we toss away without a thought, treasures are hiding.

Indeed, vegetables are not as vegetative as we might think. They are born, they live, and they die. Modestly, discreetly, perhaps since the dawn of time, they represent the most fertile encounter between nature and culture.

---

*There is a French proverb, manger des carottes rend aimable, "eating carrots makes you nice"; and in France, rather than being told that new babies are brought by a stork, children are told that they are born in the cabbage patch.

# A MATTER OF *Taste*

✢ ✢ ✢ ✢ ✢ ✢ ✢ ✢ ✢ ✢ ✢ ✢ ✢ ✢ ✢ ✢ ✢ ✢ ✢ ✢ ✢ ✢ ✢ ✢ ✢

Nothing is more variable than taste, which is dictated by the period, the place, and the social group to which we belong, not to mention the subjective and personal aspects that enter into the equation. Certain vegetables, such as the tomato, took centuries to be accepted; others, such as the parsnip, gradually lost favor, at least in France, but then reappeared with the vogue of "forgotten" vegetables.

For centuries, the word "taste" has simply designated one of the five senses, the one that enables us to perceive flavors. But its etymology, from the Latin *gustus*, actually suggests a different idea: the Indo-European root of the Saxon word *kausjar* means "to choose." To place a few markers in the history of taste is thus to reveal the choices of a society or a social group—without neglecting the thousands of nuances introduced by individuals. Our tastes define us. But there is no taste without distaste. And the history of distaste might be just as instructive: tell me what you can't swallow, and I will tell you who you are!

The history of taste in France is rooted in the fourteenth century, when the first cookbooks appeared, in particular *Le Mesnagier de Paris* (1393) and Taillevent's *Le Viandier* (1370). *Le Mesnagier* is written as a collection of advice of all kinds addressed by an old Parisian bourgeois to his fifteen-year-old bride. Moral and religious considerations precede cooking instructions and advice for keeping house or buying fresh milk at the market. This home economics manual gives us a wealth of practical information about daily life at the end of the fourteenth century, and priceless specifics about the menus and recipes of the bourgeoisie of that time.

*Le Viandier* reflects aristocratic tastes and practices, for its author, Guillaume Tirel, or Taillevent, was the master chef of kings Charles V and Charles VI. The coats of arms engraved on his tombstone, along with other hints, suggest he had been an alchemist. A transformation through flames in a quest for quintessence: cooking touches the very secrets of substance and life.

Medieval taste is characterized by a passion for color (green, yellow, red, and blue) and for spices, added "in great abundance" to all sorts of preparations. The common belief that spices served to hide the rotten smell of meat has been debunked by historians, since meat in the Middle Ages was eaten fresh and was first blanched. Reading the recipes, we notice above all that spices were added to all dishes: pies, soups, and meats. They were believed to have therapeutic virtues; the medicinal dimension of food was at the forefront in the Middle Ages. All the same, the wide latitude left to the cook and the lack of clarity in the instructions show that the addition of spices was above all a matter of taste. Spices also belonged to the realm of the imagination: they perfumed the Garden of Eden; they spoke of distant lands, of oriental landscapes that inspired dreams—such as the seed of paradise, which came from West Africa—and made those who traded in them quite wealthy, giving birth to powerful economic networks. Indeed, the exchange value of spices is the source of the French expression *payer en espèces*—literally "to pay in spices" or "in kind," but now meaning "to pay in cash."

In addition to these spices (ginger, pepper, saffron, cinnamon, galangal, cardamom, anise, cumin, etc.), acidic flavors were dominant. A large number of dishes included marinades, sauces made with verjuice (the juice of green, unripe grapes) or vinegar. An acidic flavor became more popular in France than a sweet one, and the sweet was increasingly separated from the savory. Elsewhere in Europe, however, the same dish could have combined them, or two versions of a dish existed—one sweet and one savory—to offer a choice. The acidic goes well with a largely nonfat cuisine, in which bacon, lard, or oil, depending on the region, are used as fats, butter being reserved for peasants. What little thickening was done usually involved bread or crushed almonds.

The Middle Ages borrowed "the great chain of being" from antiquity. This was a pyramidal reading of creation. The pyramid goes from inanimate elements to God, and from the low to the high. Thus animals rank above plants, and, within plants, fruits rank above leafy plants, which in turn rank above roots. Earth, water, air, and fire, in concentric circles, follow the same ascending order. This hierarchy is inscribed in a vision of the world that goes from the terrestrial to the spiritual, from hell to paradise, from the body to the soul. It was integrated, of course, into a Christian concept of the universe. Vegetables, belonging both to the inanimate and to plants, could only be considered a coarse food, good for peasants and animals.

Aristocratic tables featured marine mammals (whales, seals, sea otters, dolphins) and large birds (peacocks, swans, cranes, herons, storks, cormorants). Birds were cut up for cooking and then reconstituted, complete with feathers, for presentation. These animals lived freely, as did the nobility. The symbolism that connected the nature of the food closely to the one who ate it accounts in part for medieval tastes—and much more: I am what I eat, and I eat what I am. Beef was disdained, reserved only for bouillon; mutton, veal, and especially game, an aristocratic privilege, were preferred. The entirety, strongly hierarchized, was coherent. The microcosm and the macrocosm were in perfect harmony.

Things began to change in the Renaissance, through the influence of Italy. For Lorenzo Valla, the author of *De vero falsoque bono* (1431), taste constituted a key not only to the pleasures of the flesh but also to civilization. For the Italian nobleman Bartolomeo Sacchi, known as Platina, whose work *De honesta voluptate et valetudine* (c. 1470) was widely available in Europe in the early days of the printing press, meals represented the height of conviviality and friendship. This learned humanist in the service of the pope had at first envisioned a treatise on the pleasure of eating, but he was forced to water down his epicurism. The work was translated into French in 1505 by Didier Christol under the title *Le Livre d'honnête volupté et santé.*

✤ ✤ ✤ ✤ ✤ ✤ ✤ ✤ ✤ ✤ ✤ ✤ ✤ ✤ ✤ ✤ ✤ ✤ ✤ ✤ ✤ ✤ ✤

### PLATINA'S HERB PIE FOR MAY

Cut up and pound as much cheese as I suggested for the first and second pies. To this, when it is pounded, add juice of chard, a little marjoram, a little more sage, a bit of mint, and more parsley. When all this has been pounded in a mortar, add fifteen or sixteen beaten egg whites and half a pound of fat or fresh butter. Some also put in some leaves of parsley and marjoram, cut but not pounded, a half pound of white ginger and eight ounces of sugar. When all these are mixed together and put in a pot or well-greased pan, make them boil on coals far from the flame so they do not absorb smoke, mixing constantly until they become thick. When they are almost cooked, transfer to another pot with an undercrust and cover with an earthen pot lid until everything is cooked on a gentle fire. When they are cooked and transferred to a dish, cover with the best sugar and rose water. This dish is as much greener as it is better and more pleasing. Let Philenus Archigallus beware of this, for it digests slowly, dulls the eyes, makes obstructions, and generates stone. ✳ PLATINA, *On Right Pleasure and Good Health,* translated by Mary Ella Milham, p. 365

✢ ✢ ✢ ✢ ✢ ✢ ✢ ✢ ✢ ✢ ✢ ✢ ✢ ✢ ✢ ✢ ✢ ✢ ✢ ✢ ✢ ✢ ✢ ✢

The book gives us a very interesting glimpse into a cuisine that mixed Roman and medieval influences with pre-Renaissance innovations. Platina concludes each recipe with a nutritional and medical commentary. Thus a white tart—a sort of cheesecake—"is difficult to digest, heats the liver, causes an obstruction, causes stones and gravel and is harmful to the eyes and liver." In the book we also find, as in some ethnic cuisines, the symbolic value attached to a given part of the animal's body: thighs will make you run faster; brains, more intelligent. The important place given to aphrodisiacal virtues proves, if proof were needed, the universal link between food and sexuality. But Platina's purpose, as the title indicates, is "honest pleasure." "I understand and am speaking of that pleasure that is temperate and measured with good living which human nature desires and wishes to have."

A healthy mind in a healthy body, moderate enjoyment, harmony of the senses, a better knowledge of nature: the first book of recipes ever printed is fully inscribed in the humanist project of the Renaissance. The recipes—elderflower tart, rosemary confit with salad, or fresh beans in soup—illustrate this perfectly.

Let's remember Gargantua,* drawn away from his deplorable habits

---

*François Rabelais,* Gargantua and Pantagruel (1532).

of gluttony by his tutor Ponocrates. The meal—henceforth a wise balance between herbs, roots, and fresh water—is interspersed with readings from the ancients on the medicinal and nutritional virtues of food. For Rabelais, this change in dietary regimen was one of the symbols of the new spirit of humanism and was an integral part of a young nobleman's education.

As for Erasmus of Rotterdam, in *A Handbook on Good Manners for Children* (1530), he looks at education with a view toward civility, and in particular at the way one eats. Rules for hygiene prevail: it is forbidden to spit, to wipe your face with the tablecloth, or to make yourself vomit. It is unseemly to offer your neighbor a piece of meat you have already touched or, worse, have had in your mouth.

Table manners evolved in the direction of refinement and individualization. This latter aspect—which assumed that you had the means to buy cutlery, plates, and glasses, while the poor were still making do with trenchers, hunks of bread, rough spoons, and goblets—was extended in the Renaissance by increasingly sophisticated rules of etiquette. The fork appeared. The knife, reserved for the table, replaced the dagger or the sword. While admitting that he preferred drinking from his own glass rather than sharing one, Montaigne, after his encounter with the clerk of the kitchen of the late Cardinal of Caraffa, makes fun of the seriousness with which some speak of "palate-science":

> I put this fellow upon an account of his office: when he fell to discourse of this palate-science, with such a settled countenance and magisterial gravity, as if he had been handling some profound point of divinity. He made a learned distinction of the several sorts of appetites; of that a man has before he begins to eat, and of those after the second and third service; the means simply to satisfy the first, and then to raise and actuate the other two; the ordering of the sauces, first in general, and then proceeded to the qualities of the ingredients and their effects; the differences of salads according to their seasons, those which ought to be served up hot, and which cold; the manner of their garnishment and decoration to render them acceptable to the eye. After which he entered upon the order of the whole service, full of weighty and important considerations . . .

... and all this set out with lofty and magnificent words, the very same we make use of when we discourse of the government of an empire.*

In the sixteenth and seventeenth centuries, the taste for vegetables began to develop. The status of grains and peas diminished in favor of mushrooms, asparagus tips, hops sprouts or grapevine tendrils, artichokes, and cardoons. The chef La Varenne recommended "a thousand sorts of vegetables that are found in abundance in the countryside." There was a weakening in the medieval "great chain of being," perhaps, but court fashions were also influenced by the wars in Italy and by the queens Catherine and Marie de Médicis. In French cuisine, the role of spices diminished, with the exception of pepper, cloves, nutmeg, and cinnamon in sweet dishes. French people who traveled in Europe complained of the overspiced cooking at German or Spanish banquets, and it took all of Montaigne's curiosity and tolerance to "join the tables thickest with foreigners." Having become more common, and less costly, spices no longer really served to mark a distinction between aristocratic and bourgeois tables. Pâtés decorated with animals' heads in the medieval style now seemed vulgar. Other criteria then appeared, which followed an eternal schema clinging to the goal of social differentiation. In a century when the bourgeoisie were mimicking the nobility, it was necessary to find new codes to distinguish the two classes. Each age invents its own ways of doing this.

In the seventeenth century, "distinction through taste," to borrow the expression of the historian Jean-Louis Flandrin, occurred through a quest for the natural. Spices were replaced by herbs, chervil, tarragon, bouquet garni; rapid cooking, with the various ingredients separated, was now preferred.

"The best and healthiest way to eat roast beef is to devour it right off of the flames in its natural juices, and not completely cooked, without taking so many unnecessary precautions, which destroy through

---

*Michel de Montaigne, "Of the Vanity of Words" (1850), trans. Charles Cotton, drawn from Quotidiana, ed. Patrick Madden, April 2, 2007, and May 31, 2009, http://essays.quotidiana. org/montaigne/vanity_of_words.

their foreign ways the true taste of things," prescribed L.S.R., the anonymous author of *L'Art de bien traiter* (1674) and the paragon of absolute taste. That very talented cook, a fine technician, proved to be scornful of the "ignorant plebe." Taste was not a concern for the "commoner."

The separation of sweet and savory became more pronounced. Finally, with butter allowed in all the countries of the Reformation, the papacy, out of fear of Protestant contagion, authorized its use during Lent, favoring a fattier, less acidic cuisine. Beef made its appearance on aristocratic tables, but the "noble" cuts were distinguished from the baser ones. For Jean-Louis Flandrin, the beginnings of the gastronomic adventure began in the middle of the seventeenth century. Cookbooks set the tone: great French cuisine emerged from Pierre François La Varenne's *Le Cuisiner français* (1651), L.S.R.'s *L'Art de bien traiter* (1674), and François Massialot's *Le Cuisinier royal et bourgeois* (1691).

This tendency toward purity and the natural—which didn't come without rigorous rules—occurred at exactly the same time as the passage from the Baroque to classicism, as if that powerful aesthetic current had tapped not only into literary and artistic domains but also into daily life.

Thus the gardener Nicolas de Bonnefons, in the preface to his *Délices de la campagne* (1654), gives the following advice:

> You must try as much as you can to diversify and distinguish, by flavor and by form, the food you are preparing. A healthy soup should be a good bourgeois soup, well stocked with good, well-chosen meats, and reduced to a small amount of bouillon, without cut-up meat, mushrooms, and spices, nor other ingredients, for it should be simple because it is called healthy. A cabbage soup should taste entirely of cabbage, a leek soup of leeks, a turnip soup of turnips, and the same for the others. . . . What I say about soups I intend to be universal and serve as a rule for everything that is eaten.

————

For cooking, as for art, one must henceforth "paint from nature" (to use Molière's words), which demanded the greatest talent. "Nature" as the seventeenth century understood it, was rethought, recomposed, organized, codified, aestheticized: "nature" was a formal French garden, not an untamed forest. What was sought, therefore, was a form of truth rather than mere authenticity.

But the distance from theory to practice can be long. Just as the Baroque flowed into classicism, these culinary tendencies went hand in hand with the presentation of dishes, with decorum and table manners that were increasingly refined, even theatrical when it came to court banquets. There was an excess of luxury products (foie gras, oysters, morels, truffles), elaborate dishes, with dessert the crowning glory. In this complex gastronomy, far from the simplicity urged by Nicolas de Bonnefons, in L.S.R.'s boneless capon pie or extraordinary crawfish soup, it was at the service of flavor that bases, thickening, and juices were developed—which did not prevent the care given to a very sophisticated presentation.

———

In *Le Repas ridicule*, "The Ridiculous Meal," Boileau creates a pitiless satire of this mixture of "special" effects, medieval remnants, and nouvelle cuisine, taking aim specifically at the court caterer Mignot, who, incidentally, tried to sue the poet. Inspired by satire 8 in book 2 of the Latin poet Horace's *Satires*, and by the sixteenth-century French satirist Mathurin Régnier, Boileau lampoons a banquet that is both elegant and pretentious, where culinary bad taste combines with sins of literary taste, occasioning general mayhem with delightful burlesque effect.

✜ ✜ ✜ ✜ ✜ ✜ ✜ ✜ ✜ ✜ ✜ ✜ ✜ ✜ ✜ ✜ ✜ ✜ ✜ ✜ ✜ ✜ ✜ ✜

*Fundanius, a comic poet and a friend of Horace's, describes a dinner party given for Maecenas and his friends by the rich parvenu Nasidienus Rufus.*

How did you enjoy your swell party chez Násidiénus?
Yesterday I was trying to get you to dine with me, but was told
you'd been drinking there since midday. "I've never had such a time
in all my life." Well tell us then, if you've no objection,
What was the first dish to appease your raging bellies?

"First there was a boar from Lucania, which our gracious host kept
    telling us

was caught in a soft southerly breeze. It was garnished with things
that stimulate a jaded appetite—lettuces, spicey turnips,
radishes, skirret, fish-pickle, and the lees of Coan wine.
When this was cleared away, a boy in a brief tunic
wiped the maple table with a crimson cloth, while another
swept up scraps and anything else that might annoy
the guests. Then, like an Attic maiden bearing the holy
emblems of Ceres, in came the dark Hydaspes carrying
Caecuban wine, followed by Alcon with unsalted Chian.
Then his lordship said, 'If you prefer Falernian or Alban
to what has been served, Maecenas, we do have both varieties.'"

It's a terrible thing to have money! But do tell me Fundanius,
who were your fellow guests on this magnificent occasion?
. . .
"Our host had Nomentanus above him and Hogg below.
The latter amused us by swallowing whole cakes at a time.
Nomentanus was deputed to point out features that might have
escaped attention. For the uninitiated mob (that is,
the rest of us) were eating fowl, oysters and fish that contained
a flavour totally different from anything we had known before.
This became clear early on when Nomentanus offered me
fillets of plaice and turbot which I hadn't previously tasted.
He then informed me that the apples were red because
    they'd been picked
by a waning moon. If you wonder what difference that makes,
    you'd better
ask the man himself."

∗ *The Satires of Horace and Persius,* translated by Niall Rudd (Penguin Classics, 1973), pp. 122–23

✤ ✤ ✤ ✤ ✤ ✤ ✤ ✤ ✤ ✤ ✤ ✤ ✤ ✤ ✤ ✤ ✤ ✤ ✤ ✤ ✤ ✤

Gradually, a certain number of arbitrary codes appeared whose sole function was to distinguish people by good manners. Eating with one's knife and cutting bread rather than breaking it apart were stigmatized actions reserved for country folk. Finally, the order of the courses, the seating at table, became more complex and contributed to effecting separation between masters and valets, the poor and the rich, aristocrats and bourgeois, princes and minor marquis.

The diversity of flavors, maintained by a multiplicity of dishes, not all of which, of course, the guests were expected to eat, imperceptibly gave way to a debate among specialists on "good taste."

And it was precisely at that time that an extension to the term "taste" appeared. "You have good taste," one of Molière's *Précieuses ridicules* proudly says to herself. For Voltaire, "just as a physical bad taste consists in being pleased only with high seasoning and curious dishes, so a bad taste in the arts is pleased only with studied ornament, and feels not the pure beauty of nature."* Good taste assumed simplicity, discretion, refinement, elegance, harmony; it shunned excess, ostentation, sophistication, exuberance. Measure, balance, truth. Good taste then would concern what is felt, intuition, instinct almost as much as education.

In the eighteenth century, the word "taste" took on a specifically positive meaning: a man of taste was henceforth a man of *good* taste.

"There are a thousand people with good sense for one man of taste," remarked Diderot in his *Lettre sur les aveugles*, "and a thousand people of taste for one of exquisite taste." "Exquisite" became the supreme form of taste.

By going from the concrete to the figurative, taste had definitively chosen its camp. Gastronomes of the nineteenth century such as Brillat-Savarin and Grimod de La Reynière took it upon themselves, often with remarkable literary talent, to codify the habits, tendencies, and preferences of their time. As for the debate between partisans of original flavor and fans of knowledgeable masking, between an aspiration to classicism and a Baroque exuberance, between simplicity and technicality— it continues even today. It is the essence of any history of gastronomy.

---

*A Philosophical Dictionary from the French of M. de Voltaire (*London, 1843*), 2:523.

# Some Vegetable Histories

THE

# *Cardoon* AND THE *Artichoke*

❖ ❖ ❖ ❖ ❖ ❖ ❖ ❖ ❖ ❖ ❖ ❖ ❖ ❖ ❖ ❖ ❖ ❖ ❖ ❖ ❖ ❖ ❖ ❖

Let's think for a moment: what distinguishes a donkey from a man?

From a thistle to a cardoon, from a cardoon to an artichoke, from a wild plant to a cultivated plant and then to a plant developed by means of research, this long path through hundreds of years may provide an answer.

---

It all began with thistle heads, or capitula, from which prehistoric peoples extracted the edible and very delicately flavored heart. Although we still don't know exactly when the thistle was domesticated, the tradition of eating it continued for a long time in Italy. In 1787 Goethe ran into two men in Sicily who, with obvious relish, were eating thistles they had picked on the roadside: "To our amazement, we saw these two dignified gentlemen standing in front of a clump of thistles and cutting off the tops with their sharp pocket knives. Carefully holding their prickly acquisitions by the finger tips, they pared the stalk and consumed the inner portion with great gusto, an operation which took them some time."[*]

The same custom is found on the other side of the Mediterranean, in Syria, with *akkoub*, and in the Maghreb with another thistle flower of which the heart and ribs are eaten. In the third century BC, the Greek Theophrastus, in his *Enquiry into Plants*, mentions *cactos*, cardoon, and

---

[*]*Johann Wolfgang von Goethe, Italian Journey, trans. W. H. Auden and Elizabeth Mayer, p. 277.*

*skalias*, the heart of the capitulum. Columella, a Roman agronomer from Cadiz, gives advice in his *De re rustica* on how to obtain fine capitula of *cinara*; and Pliny, in his *Natural History* in the first century AD, informs us that *carduus*, ancestor of the cardoon, was grown not only in Sicily and around Carthage (in present-day Tunisia), but also in Cordoba in Andalusia, where it was an important source of revenue. Cadiz and Cordoba had been Phoenician and then Carthaginian trading centers before they became Roman. The Roman chef Apicius offers recipes for cooking the stalks and buds as well as the hearts of *carduus*. For example, he suggests sprinkling the boiled hearts with a cornstarch sauce made with crushed celery seeds, rue (a meadow plant), honey, pepper with light wine, garum (fish-based sauce), and olive oil, which is mixed with the cornstarch before being sprinkled with pepper. There is no doubt that cardoons were thoroughly enjoyed throughout the Mediterranean territories of the Roman Empire.

And then they disappeared for a thousand years.

How did cardoons return? Actually, they continued to be eaten in Tunisia and also in Andalusia, where they later benefited from new cultivation techniques thanks to Arab expansion in the eighth century. Gardeners attempted to develop the flower, the capitulum, on the one hand, and the leaf stalk and central vein, on the other. Gradually, the artichoke and the cardoon became distinct from each other.

It was the cardoon that first traveled from Andalusia to Sicily (1300). The artichoke, which was rarer and more delicate, made its entrance into Italy a century later.

The cardoon seems to have arrived in France directly from Spain and traveled up the Rhône Valley as far as Lyon, "a true land of thistles," according to the sixteenth-century agronomer Olivier de Serres. It is still one of the specialties of Lyon and the Dauphiné, where it is blanched in cellars. But it wasn't until the end of the seventeenth century that this giant plant, which could reach a height of eight or nine feet, was tamed: "The thistle, no matter how it is handled, is still covered with strong sharp thorns," noted the same Olivier de Serres. It then reached Île-de-France and traveled across the English Channel, though in England it was considered only an ornamental plant, unlike its cousin the artichoke, which was thoroughly enjoyed as a food.

Was the artichoke introduced directly from Navarre into Naples, following the annexation of Naples by the Spanish crown? Or did it come through Sicily? The lexicon seems to lean in favor of the first hypothesis, as the Neapolitan *carcioffola* resembles the Spanish *alcachofa*. In any event, the artichoke then traveled to Tuscany, introduced by Filippo Strozzi in 1466; then toward Nîmes and the Vaucluse region, where in the sixteenth century it began to be grown like other vegetables of Mediterranean origin, such as melons and Swiss chard. So it was not Catherine de Médicis who imported it into France, as has been claimed, though she did contribute to promoting a vegetable she adored, encouraging people to eat it. At the time it was a delicacy, a refined dish. The figure that incarnates *Summer*, in a painting by the sixteenth-century Italian mannerist Arcimboldo, is wearing an artichoke on his lapel, like a flower or a piece of jewelry. The seventeenth-century engraver Abraham Bosse chose it to illustrate taste. But it wasn't until the eighteenth century, once the usual resistance to a new food was overcome, that it was widely recognized as one of the most delicious vegetables.

––––––––

Language reflects these botanical and gastronomic adventures: the Latin *carduus* gave us the French *carde*, *cardon*, and the English "cardoon." The *carde* is the fleshy edge of the cardoon. It is also the name that in the thirteenth century was given to the head of the thistle with which one *carded* wool to untangle it.

The artichoke derives its name from the Arab word *al-harcharf*, which became *alcachofa* in Spanish and *carciofo* in Tuscan; the Lombard *articcioco* led to the French *artichaut* and the English "artichoke."

The scientific names underscore both vegetables' kinship with the thistle: *Cynara cardunculus* (cardoon) and *Cynara scolymus* (artichoke), according to Linnaeus's classification in the eighteenth century.

In the sixteenth century, the cardoon and the artichoke were considered fruits or side dishes, and Rabelais includes them in the feast of the Gastrolâtres in between peaches and cakes ("Basket peaches, artichokes, pastries, cardoons, sweets, beignets"). "Artichokes both green and purple or red" were grown by La Quintinie in the King's Vegetable Garden at Versailles, as were "artichoke stalks" and "cardoons from Spain." But the cardoon attained its greatest popularity in the nineteenth century. In 1888 it was served in the Spanish style at the royal table in Belgium. For Grimod de La Reynière, the cardoon was "the apex of human knowledge, and a cook who can make an exquisite dish from cardoons may call himself the leading artist in Europe." The gastronomer advises gourmands to serve them with sauce, coulis, or marrow; ungarnished, or served with parmesan, it was quite good enough for working people or those who lacked taste.

Marrow was a way of making it more bourgeois—even of ennobling it. A vegetable of humble origins, a distant relative of the thistle? Indeed, Françoise, Aunt Léonie's incomparable cook, at the house at which the narrator of *Remembrance of Things Past* spends his vacations, would prepare cardoons with marrow for the Sunday guests:

> For upon the permanent foundation of eggs, cutlets, potatoes, preserves, and biscuits, whose appearance on the table she no longer announced to us, Françoise would add—as the labor of fields and orchards, the harvest of the tides, the luck of the markets, the kindness of neighbors, and her own genius might provide; and so effectively that our bill of fare, like the quatrefoils that were carved on the porches of cathedrals in the thirteenth century, reflected to some extent the march of the seasons and the incidents of human life—a brill, because the fish-woman had guaranteed its freshness; a turkey, because she had seen a beauty in the market at Roussainville-le-Pin; cardoons with marrow, because she had never done them for us in that way before.*

Cardoons with marrow thus appeared in Combray, a village in the Beauce, as a Sunday treat, a culinary delight worthy perhaps of the Guermantes.

———

But this didn't prevent the gradual decline of the cardoon, except in the area around Lyon, where a cardoon festival is held every December in Vaux-en-Velin. The history of vegetables, however, has many shoots in its pot: we see the cardoon returning to European markets thanks to the populations from North Africa who have never stopped eating it, especially in Moroccan couscous broth or Tunisian dishes. The same is not true of the artichoke, which French consumers have been avoiding

---

*Marcel Proust, Swann's Way, trans. C. K. Scott Moncrieff (2002), p. 60.

lately—unless a taste for them returns via Italy, the world's foremost producer and consumer of artichokes.

———————

What, then, distinguishes a donkey from a man? The French say "dumb enough to eat straw"—the straw we pull off the artichoke heart, perhaps?—but also "dumb enough to eat cardoons." Is it because a donkey doesn't know how to pull the thorns out of life to reach the delicious heart of things? In medieval symbolism the cardoon represents the trials of existence, or the suffering of a martyr, as illustrated by Christ's crown of thorns or thistles; but it is also believed to make women fertile and men virile. It becomes the symbol of Lorraine and Scotland, and delivers the message that "whoever touches gets pricked," the symbol of armed resistance.

The artichoke, once evolved, in addition to its depurative and diuretic virtues (as confirmed by modern science and due to cynarine), is believed to have acquired aphrodisiacal properties, which contributed largely to its success in the Renaissance and later, if we are to believe a recipe lovingly devised by Mme du Barry for her illustrious lover Louis XV: venison and pheasant cooked in white wine and accompanied by artichokes and asparagus with plenty of pepper. In French, *avoir un coeur d'artichaut*, "to have an artichoke's heart," is used to describe someone fickle, who falls in love easily and often. And the leaves you pull off with your fingers, dip into the sauce, and suck may indeed foretell other pleasures.

This popular song illustrates the belief in the rousing virtues of the artichoke:

*Colin mangeant des artichaux*
*Dit à sa femme: Ma mignonne*
*Goûtes-en, ils sont tout nouveaux;*
*... La belle avec un doux maintien*
*Lui dit: Mange-les toi que mon coeur aime*
*Car ils me feront plus de bien*
*Que si je les mangeois moi-même.**

———————————————————————————————

*Colin, eating some artichokes / Says to his wife: My dear / Taste them, they are very fresh; / ...
The sweet-looking lady / Says to him: Eat them yourself, my love / For they will do me more
good / Than if I eat them myself.

Or this very telling cry from a Parisian market:

*Artichauts! artichauts!*
*C'est pour Monsieur et pour Madame*
*Réchauffer le coeur et l'âme*
*C'est avoir le cul chaud.*\*

It's not surprising, then, that the artichoke was Sigmund Freud's favorite plant. His wife often brought him some artichokes from the market. It reminded him of something from his childhood: a book that his father had given him for his fifth birthday that he had happily torn apart with his sister! How can I forget, Freud notes, "the infinite joy with which we tore the pages from that book (page by page, as if it had been an *artichoke*)?" Where would the appetite for knowledge be hiding? Later he developed a true passion for books.

———

What, again, distinguishes us from donkeys? Donkeys are content with thistles; we prefer artichokes.

---

\**Artichokes! artichokes! / They're for all the ladies and gents / To heat up their hearts and souls / Your ass will be ready to go!*

THE

# *Jerusalem Artichoke*

✤ ✤ ✤ ✤ ✤ ✤ ✤ ✤ ✤ ✤ ✤ ✤ ✤ ✤ ✤ ✤ ✤ ✤ ✤ ✤ ✤ ✤ ✤

As we have seen, vegetables have gone through countless unexpected adventures throughout the centuries, and their history is sometimes as eventful as that of humans. The Jerusalem artichoke provides a fine example. Cortez landed in Mexico in 1519; Pizarro, in Peru some fifteen years later; and Jacques Cartier, choosing to travel the northern route, entered Canada in 1536. And at the beginning of that century, Christopher Columbus traveled along the coasts of Brazil. Information spread slowly to the general public and was sometimes so fantastic that those who heard it were perplexed—or amazed. They were fascinated by the stories of distant lands and their bizarre inhabitants. Travel tales in the sixteenth century were extraordinarily popular, as were the exotic plants brought back by the explorers, who became conquistadors, then colonizers. People were cautious, of course, when confronted with these new plants, which they didn't yet know how to cook, but they were curious as well. Tomatoes, potatoes, yams, beans, chili peppers, bell peppers, and squash were brought back to Europe by the explorers and conquerors of the Americas.

Samuel Champlain, who came from the province of Saintonge and was born in Brouage, left for Canada in 1603 with a fur-trading expedition to establish a base there. He was the first to note those strange tubers eaten by the Montagnais, natives from the north of the Saint Lawrence River. These people belonged to the Innus tribe, who were hunter-gatherers. When he returned, Champlain published an account of his voyage: *Des Sauvages*.*

---

*The first English-language edition appeared in 1625.

Champlain journeyed twenty-one times to New France; he founded Acadia and the city of Québec. But what is important for our purposes, he was the first to mention the plant with the bizarre shape that was later called *topinambour* in French—or, in English, "Jerusalem artichoke." According to Champlain, these tubers tasted like artichokes, an impression confirmed by Marc Lescarbot, a lawyer, who spent several months in Acadia, along with Champlain and a few fellow travelers. Lescarbot's *Histoire de la Nouvelle-France*, which he published on his return, was hugely successful. Like Champlain, he described the customs and resources of the "savages" of those northern lands, still less known than those of South America: "There is also in this land a certain sort of large root that resembles a turnip or a truffle, quite excellent to eat, having a taste like a cardoon, but even more pleasant, which, when planted, multiplies as if out of spite in a marvelous way." Marc Lescarbot brought some of those roots home with him, and they did indeed reproduce incredibly fast. The Indians gave it the soft-sounding name of *chiquebi*, he specified, but he wanted to call it *canada*, in honor of its original land. Alas, they were sometimes called "earth walnuts," truffles, even potatoes or Canadian artichokes. This handful of examples conveys the difficulty of a taxonomy that was still in its infancy. The plant's botanic origins are similarly confusing.

In the meantime, the tuber traveled. It reached Holland and Germany, and in 1616 it was being grown in the gardens of Cardinal Farnese in Rome. The Italian Fabio Colonna identified a *helianthus*, believing it to have come from Peru, like the potato. Another botanist, the Swiss Gaspard Bauhin, did attribute it to the Canadians but turned it into a *chrysanthemum*. This strange object—deformed, purplish, and wrinkled—thus had neither an established name nor a well-determined origin.

During the same period, in 1613, a delegation of Tupinamba Indians from Brazil landed in France to attend the entrance of the young King Louis XIII into Rouen.

The Tupinambas (no one knew quite what to call them: Topis? Tupis? Tupinambus?) were not unknown to the educated public. Jean de Léry called them Toüoupinambaoults in his *Histoire d'un voyage fait en la terre du Brésil*, published in 1578 and elevated to the rank of "breviary of ethnology" by Claude Lévi-Strauss.

In its time, this book delighted Montaigne, who in fact met three of the Indians in 1562 on the occasion of the entrance into Rouen of young Charles IX and his mother Catherine de Médicis, an adventure that gave birth to one of the most famous passages in the *Essais*. The "cannibals" seemed very intelligent to him, which led him to wonder about barbarism. What does it mean to be a savage? "I find that there is nothing barbaric or savage in this nation, as to what I have been told about them; unless we must call 'barbarism' that which we are not accustomed to" (*Essais* I, 31).

It was a fine lesson on the relativity of customs, values, and morality, but also a reflection on the concept of nature, and the beginning of the myth of the noble savage. Consequently, when the Tupinambas landed in France in 1613, they were immensely popular. People clamored to see them dance and shake their feathers, and shivered to think of their culinary practices, which consisted of roasting their enemies after cutting off their arms and legs. That custom had also inspired some reflections in Montaigne, for whom "there is more barbarity in eating a living man than to eat him dead, to tear apart through torment and anguish a body still full of feelings, to roast him bit by bit, and have him bitten and eaten alive by dogs and pigs (as we have not only read but seen in recent memory, not among ancient enemies, but between neighbors and co-citizens, and, which is worse, under the pretext of piety and religion)." Or how to denounce torture and wars of religion under the guise of anthropology.

Through a strange sleight of hand, the Canadian *chiquelis* found their way: in French they became *topinambaux* (Lescarbot's term), then *topinambous*, and finally *topinambours*. The linguistic wanderings of the *Helianthus tuberosus* were not over, however: for the English, it became the "Jerusalem artichoke," an unauthorized label that resulted from the alteration of the Italian *girasole* (sunflower) pronounced with an English accent and a flavor that resembled the artichoke.

The vegetable began to create quite a stir and was even served at the royal table. But, perhaps due to the pejorative connotation of the notion of savage in the seventeenth century, the word *topinamboux* gradually began to be applied to vulgar and limited people, as seen in these lines by Boileau:

*J'ai traité de Topinamboux*
*Tous ces beaux censeurs je l'avoue,*

*Qui, de l'antiquité si follement jaloux*
*Aiment tout ce qu'on hait, blâment tout ce qu'on loue.*\*

Clearly quite fond of the metaphor, Boileau also called the Académie française *topinamboue*. This semantic evolution was a bad sign.

"Regarding the *topinambours*," notes the doctor Guy Patin, dean of the faculty of medicine in Paris, in a letter of October 29, 1658, "it is a plant that comes from America, for which there is no use in Paris, nor elsewhere, as I have been told; in the past, gardeners would sell the root, which is bulbous and tubular; but this was not continued; it required a lot of salt, pepper, and butter, which are three undesirable things."

The popularity of the Jerusalem artichoke didn't last long. By the end of the seventeenth century, its number was up. For Furetière, in 1690, it was a "round root, which comes in knots, which the poor eat cooked with salt, butter, and vinegar." Rustic and abundant, the Jerusalem artichoke lost its exotic aura. De Combles, the author of *L'École du jardin potager* (1752), minces no words: he denounces the Jerusalem artichoke as "the worst of vegetables, but still eaten by the people."

It was eaten during Lent, as a sign of penance. For Jaubert, one of the editors of the *Encyclopédie*, "these roots are tasteless, watery, insipid, very gassy and unhealthy, which makes them ignored almost everywhere in 1771." It was believed that the Jerusalem artichoke, like the potato a little later, caused leprosy. What is unusual is that this suspicion came after a first phase of great popularity, and not when the vegetable was a novelty.

The Jerusalem artichoke, then, was already essentially dethroned when the potato arrived. It was used as a fodder crop. In some parts of Normandy, it was even used for pig feed, as validated in the 1830 Annuaire de Falaise.

It wasn't until World War II that people rediscovered the charms of the Jerusalem artichoke, which became inseparable from the chubby rutabaga—the Laurel and Hardy of the Occupation. Very prolific, the two were substituted when there was a lack of potatoes due to frosts, the condition of the roads, transportation problems linked to gas ra-

---

\**I have given the name of Topinamboux / To all those nice censors, I admit / Who, so in-sanely jealous of antiquity, / Love everything we hate, and criticize everything we praise.*

tioning, and requisitions by persons derided as "potato beetles." In February 1942, however, the Jerusalem artichoke still failed to appear in the markets; farmers kept them for their animals, speculating on an increase in their price. "The customers who disdained them last winter will be happy to have them!" noted a prefectorial report.

For better or for worse, the Jerusalem artichoke replaced the potato for a while. Its purgative virtues earned it fifty years of purgatory following the end of World War II. People had eaten more than enough Jerusalem artichokes and rutabagas during the war. It was out of the question to serve them again.

It wasn't for another fifty years that the Jerusalem artichoke would be rediscovered in the basket of forgotten vegetables that had fallen out of style but then had acquired a new, very chic life as "vintage" foods. In the last few years, it has reappeared on the menus of great chefs, with foie gras, fresh truffles, or as an accompaniment to trout—far from pig food! And today this vegetable is found just about everywhere. An apotheosis before another disappearance? Only the future can tell. Thus the great scales of the history of food, and of history itself, tip now to one side, now to the other. (The potato was finally adopted in France to fill the dearth of vegetables during famines, its virtues discovered by Parmentier when he was a prisoner in Prussia during the Seven Years' War.)

But the Jerusalem artichoke has one card up its sleeve. After being fermented and distilled, it can actually provide a form of fuel, already used in 1943 as ersatz gasoline, and more recently as a biocarburant in its adoptive land of Brazil.

# *Cabbage*

✢ ✢ ✢ ✢ ✢ ✢ ✢ ✢ ✢ ✢ ✢ ✢ ✢ ✢ ✢ ✢ ✢ ✢ ✢ ✢ ✢ ✢ ✢ ✢ ✢ ✢

After the cardoon and the Jerusalem artichoke, we now enter the familiar world of cabbage. Finally, a vegetable that we have all known forever. Because that's what cabbage is: a concentrate of affective memory, a substantial food for the body, but also a vegetable that speaks to us of a territory, a country, a past, a history, a vegetable that has made its way into our everyday language.

All cabbages—and there are many—belong to the Brassicaceae family (formerly Cruciferae), one of the most diverse families. The genus *Brassica* derives its name from the Greek *prasikē*, "vegetable." Since early antiquity, cabbage has reigned as the vegetable par excellence. The French *chou* comes from the Latin *caulis*, "stalk," no doubt an allusion to the tall stalk of wild cabbage. Delicate or hearty cabbages (white or red cabbage, or green cabbages from Milan), cauliflower and romanesco cabbage, broccoli, Brussels sprouts, kohlrabi, turnip-cabbage, or rutabagas: countless varieties of European cabbages, joined by their Chinese cousins, *pei tsai* and *pak choy*, found in their country of origin since the fifth century.

This diversity testifies to the versatility of the species, its ancient domestication, and the widespread interest it has aroused. Nor should we forget the varieties of wild or semi-wild cabbages, often imported into a very circumscribed area.

Most of our cabbages have descended from distant ancestors, which grew freely on the European coastline. You can still find them on the cliffs of the English Channel—wild cabbages with yellow flowers that resemble cultivated fodder cabbages. Only cauliflower and broccoli are native to the Mediterranean area.

Cabbage is one of our oldest vegetables, already present in the Paleolithic, where it was gathered on riverbanks. It was cultivated in the Neolithic, around seven thousand years ago, one of the first plants to have been domesticated. Cabbage seeds have even been found in lakeside caves in Switzerland, far from its original habitat. Also very early on, cabbage was cultivated and eaten in quite varied forms. Today every country produces its specialty: cauliflower in France, hearty cabbage for sauerkraut in Germany, broccoli in Italy, Brussels sprouts in the United Kingdom. Chinese cabbage—of which the largest producers of course are China, Japan, and India—is grown widely in Spain, Denmark, and the Netherlands.

Brussels sprouts are descendants of fodder cabbage, not primitive cabbage. Originally from Italy but cultivated in Belgium in the thirteenth century, they were grown in the dried marshland of the commune of Saint-Gilles, following the construction of the second ramparts of Brussels.

Cauliflower made its entry into Western Europe at the end of the fifteenth century, under the name of Syrian cabbage or Cypriot cabbage. Under Henry IV, however, it kept its Italian name *cauli-fiori*, similar to the English "cauliflower." Still rare at the time, it was considered a delicacy until the eighteenth century, and it wasn't widely cultivated in France until around 1830. Today it is the vegetable most commonly grown in France. Have you ever seen a bouquet of cauliflower up close? Or, better yet, a romanesco cabbage? It is a fractal: each flower is itself composed of smaller flowers, and so on ad infinitum.

Broccoli (from *broccolo*, "shoot," in Italian), originally from the eastern Mediterranean, has been cultivated over the ages in Italy. In France it was long confused with cauliflower or the buds of green cabbage. Broccoli spread throughout the world following Italian emigrants who left to make their fortunes in the United States at the beginning of the twentieth century, and it returned to Europe in the 1980s. Its color varies: it is white in England (possibly a cross between cauliflower and broccoli), green in Italy and Switzerland, and purple in Sicily, where it originated.

Here it's worth mentioning the rutabaga, which saw the light of day in Scandinavia in the Middle Ages by means of crossing a cabbage with a turnip. Its name comes from the Swedish *rotabaggar*.

The French word for soup, *potage*, encompasses everything that is brought to a boil in a pot with water, vegetables, and/or meat (the Roman *olus*). The later word, *soupe*, derived from the Germanic *suppa* (modern German *Suppe*), once designated the slice of bread onto which the soup was poured (the reverse of dunking), but then came to mean the soup itself. Every gastronomer will tell you his or her version—the only true one—of the distinction between the words *soupe* and *potage*. Is *soupe* thicker? Less refined? This nuance is recent. In *Le Repas ridicule*, for example, Boileau uses both interchangeably. *Bouillon, brouet, soupe, potage, consommé.* I personally prefer Joseph Delteil's definition: "*Soupe* is the same as the hare: there is only one in the world." And he adds, magnanimously: "One dish in three persons: in spring it is bean *soupe*; in summer, *soupe* with green beans; in winter, cabbage *soupe*" (*La Cuisine paléolithique*).

Everything suggests that cabbage soup is one of the most ancient of dishes. It is found in various forms in most of Europe, including the south of France, as is seen in Louis Stouff's study* of food and supplies in fourteenth- and fifteenth-century Provence: for many months, cabbage soup, along with bread, was a basic food for the people of Provence, just as borscht was for Russian peasants. In medieval educational establishments, it was served to the boarders more than half the year.

In the Middle Ages, indeed, cabbage was a staple for peasants throughout Europe. It was easy both to grow and to store. It grew in vegetable gardens, which weren't taxed. In Normandy, cabbage cultivation on rented farms was regulated by tradition. Local custom in the region of Avranches, for example, as late as 1930 specified that one-quarter of a garden must be reserved for common cabbage, one-quarter for hearty cabbage, one-quarter for peas, and one-quarter for various other vegetables; so half of the garden was devoted to cabbages. I know a number of villages called Les Choux ("The Cabbages"), and you can smell the sauerkraut in the name of one Alsatian town, Krautergersheim.

In the countryside, cabbage soup was eaten daily, occasionally more than once a day, sometimes thickened with turnips, beans, or other root vegetables, at other times—in the richest homes—accompanied by

---

*Louis Stouff*, Le Ravitaillement et l'Alimentation en Provence au XIV<sup>e</sup> et XV<sup>e</sup> siècle (1970).

bacon or meat. It would thus become the symbol of peasant culture well beyond the Middle Ages: nutritious and hearty, it warms you up in cold weather, fills you up when you're hungry.

We find an example in Eugène Le Roy's novel *Jacquou le Croquant*, set in early nineteenth-century Dordogne. Jacquou, famished, is welcomed in by the village priest. The priest's servant fills a bowl with cabbage soup, which the child devours, still standing, at the end of the table. When he has finished, the servant pours him a little red wine, which he drinks before eating a second bowlful of soup, more calmly, seated in front of the priest.

This comforting soup scene is the counterpoint to another scene at the beginning of the novel, when Jacquou and his mother return to their icy hovel after escaping from a wolf: the entire meal that Jacquou's mother sets down in front of him consists of "a bowl of corn flour dissolved in water, cooked with cabbage leaves, without a scrap of bacon in it, and quite cold." Cabbage soup could represent abundance and even a certain refinement. Invited to the table of some rich peasants, Jacquou discovers their table manners:

> The soup was poured, we sat down, and the old woman served each person a dish filled with good cabbage and bean soup. I was surprised to see Duclaud eat his soup with his spoon and fork at the same time. At home we didn't use that method, for the good reason that we didn't have any forks. When we had a stew of potatoes or beans for supper, we ate it with spoons. For meat, we used a knife and our fingers; but that was only once a year, at Carnival time.

Cabbage soup also exemplified food at its most basic in a text from Émile Zola's *Le Ventre de Paris*, in which the seller of cabbage soup and her customers manifest the essence of early humankind, still close to animals in their needs and fears:

> At one corner a large circle of customers had gathered round a vendor of cabbage soup. The enameled tin cauldron full of broth was steaming over a small brazier, through the holes of which could be seen the pale glow of the embers. The woman, armed with a ladle, took some thin

slices of bread out of a basket lined with a cloth and dipped some yellow cups into the soup. She was surrounded by neatly dressed saleswomen, market gardeners in overalls, porters with jackets that bore the marks of the loads they had carried, poor ragged devils—in fact, all the hungry early-morning crowd of the markets, eating, scalding their mouths, and sticking out their chins to avoid staining their clothes as they lifted the spoons to their mouths. Claude blinked, delighted with the sight, trying to position himself so as to get the best perspective, trying to compose the scene into a satisfactory group. But the cabbage soup gave off a terrible smell. Florent turned away, put off by the sight of the full cups, which the customers emptied in silence, glancing sideways like suspicious animals.*

---

The "terrible smell" of the cabbage (due, in fact, to sulfur molecules, glucosinolates) gradually became the cabbage smell that signifies the crude nature of peasants, then simply implied poor food. Traces of this symbolism can be found in Roger Martin du Gard, as well as in George Orwell's 1984, where it illustrates all that is sordid. The strong smell of cabbage constitutes a distinctive, and negative, social symbol. We are at the very heart of the notion of taste—in both senses of the term. The truffle, too, has a strong odor. But in that case, the rarity of the product, beginning in the seventeenth century, made its smell a sign of distinction, even of luxury, whereas the odor of cabbage, a common vegetable, thought of as rural, evoked poverty or crudeness.

Gradually, cabbage soup became a prime element in caricatures of the life of peasants, depicted as vulgar and stupid.

In Jean Giraud's film *La Soupe aux choux*, two peasants, both drunk, are visited by an extraterrestrial. As a sign of hospitality, they offer it some cabbage soup, the starting point for a scene of flatulence to very good effect. The degeneration of the image of the peasant into burlesque is all the more obvious since the scene borrows a theme from ancient mythology: Philemon and Baucis, the legendary couple who became a symbol of fidelity, welcome Zeus, whom they don't recognize, and offer him "a dish of cabbage topped by a thin slice of salt pork"—in

---

*Émile Zola, The Belly of Paris, *trans. Brian Nelson (Oxford University Press, 2008), p. 23.*

other words, the best they have to offer. This legend shows how highly cabbage was regarded in antiquity, praised among the Greeks by Aristotle and Theophrastus and among the Romans by Cato the Elder.

———

Thus, cabbage, a substantial peasant food in the Middle Ages, became the emblem of a popular, inelegant food that smelled bad and had uncomfortable side effects.

It would take all of Grimod de La Reynière's gastronomic intelligence to rehabilitate the cabbage in his *Almanach des gourmands* (1803, p. 145), while nonetheless associating it with country dishes:

> Cabbage is a great help in cooking, even in sophisticated cooking. A talented artist knows how to derive an advantageous result from this vegetable, unjustly scorned by the arrogant, to vary his soups, garnishes, and side dishes. We have seen that a cut of beef, and even a mature fowl, are honored to be flanked by a thick wall of cabbage. Everything depends on the seasoning. Just as the most vulgar terms are ennobled under the pen of a great poet, a cabbage à la bavaroise, which by that name is the preferred bed of an andouille sausage, is no ordinary stew. Indeed, throughout all of Germany, and even in Alsace, they make a dish with fermented red cabbage known by the name of sauerkraut, which, shedding the cabbage of all of its negative qualities, makes it into a food that is as healthy as it is agreeable.

———

Early on, in northern regions of Europe, the question arose of how to conserve cabbage. Brine provided an answer: after cutting the hearty cabbage into strips, salt was added to promote lactic fermentation for a month or two. Conserving in brine is very ancient and practiced throughout the world, particularly in northern Europe, from Flanders to Russia. It can be found in different forms and for various foods, including turnips and beets or even meat; *picklefleish*, beef conserved in brine and thinly sliced, is a traditional Ashkenazi dish. Brine has the advantage of preserving the natural qualities of fresh cabbage, especially its vitamin C, which protects against scurvy. A diet including sauerkraut contributed to the rise of the Dutch navy in the sixteenth and seventeenth centuries. Thomas Cook also pointed out the advantages during his second trip around the world (1772–75): none of his sailors were stricken with scurvy, which was decimating crews at the time. At the beginning of the eighteenth century, sauerkraut had become "the main food of the natives of the land," noted a doctor, referring to Alsace. Sauerkraut thus proved to be a marker of national or regional identity, food and nationality being closely linked. For example, since the seventeenth century the Hungarians had made sauerkraut their national dish, although it was eaten throughout all of central Europe. Why? Because it enabled the unification of all the subjects under the Hungarian crown around a common dish, which neither language nor religion could do. But since sauerkraut existed in other regions of Europe, it didn't work as a symbol beyond the Hungarian borders. Starting in 1870, therefore, goulash, a local rural specialty made of beef and paprika, was promoted to symbolize Hungary.

✢ ✢ ✢ ✢ ✢ ✢ ✢ ✢ ✢ ✢ ✢ ✢ ✢ ✢ ✢ ✢ ✢ ✢ ✢ ✢ ✢ ✢ ✢

**MY SAUERKRAUT**

*Cooking time: 3 hours. Serves 8*

4½ lbs. unboiled cabbage
8 Strasbourg sausages
1 smoked pork shoulder
1 pork knuckle
1 rack of pork (1½ lbs.)
8–12 knockwursts
8 thin slices of smoked pork belly
¼ lb. goose fat
6 juniper berries in a small muslin bag
1 onion stuck with 2 cloves
2 cups Riesling
1 bay leaf

· Blanch the cabbage, and then rinse it several times under running water. Drain well.
· Melt the goose fat in a heavy pan. Line the bottom of the pot with the

slices of pork belly. Add the cabbage, onion, cloves, juniper berries, bay leaf, salt, and pepper. Add the Riesling and boil uncovered for 1 hour on low heat. Add the rack of pork and boil for another 1½ hours, watching the level of the juices. Add a little water if necessary.

· In the meantime, boil the smoked shoulder in another pan. When it is fully cooked, add it to the cabbage and let simmer for 30 minutes. Steam or boil some peeled potatoes.

· In a third pan, poach the pig knuckle and the sausages, taking care that they don't burst.

· Ladle the cabbage onto a very warm serving platter; arrange the meats around it and the sausages on top.

*Inhale!*

✤ ✤ ✤ ✤ ✤ ✤ ✤ ✤ ✤ ✤ ✤ ✤ ✤ ✤ ✤ ✤ ✤ ✤ ✤ ✤ ✤ ✤ ✤

And since we are in Hungary, it would be a pity not to make a detour through *stuffed cabbage*, a true cult dish in all of Eastern Europe. Stuffed cabbage feeds the soul: it is a dish that speaks to us of patient work, slow cooking, and considerable culinary skill; and it is never the same twice.

"To prepare stuffed cabbage: alchemy of cooking, happiness of heat, seduction of melting, glory of perfection, sincerity of tradition, rhetoric of invention," writes Allen S. Weiss in a little book entirely devoted to the celebration of this dish: *Reflections on Stuffed Cabbage*.

For the singer Michel Jonasz, of Hungarian origin, the odor of stuffed cabbage recalls distant Sundays with his family before everyone scattered. Cabbage is filled with nostalgia, that of a united tribe, still connected to its roots thanks to grandmother's cooking:

*Et ça sentait le chou farci*
*Ça sentait l'amour aussi*
*On avait tous le coeur au chaud*
*Comme la soupe sur le réchaud.**

---

* And it smelled like stuffed cabbage / It smelled like love, as well / All our hearts were warm / Like the soup on the warmer.

The fashion designer Paul Poiret, in 1929, offered a recipe that is as short as it is surprising: "Cabbage with jam: take a small leaf of very white and very curly cabbage. Spread it with good gooseberry jam. Eat!"* And Serge Gainsbourg, whose father was Russian, turns cabbage into a sort of self-portrait hovering between self-derision and tragedy:

*Je suis l'homme à tête de chou*
*Moitié légume moitié mec.*†

There is a lot of wordplay here, for *le chou* in Parisian argot is also "the head," which is perfectly logical since the *cabus*, the smooth hearty cabbage, is known as *capu*, or "cabbage-head," and probably derives from *caput*, "head." Years ago, *piquer un cabus* (literally, "stabbing a cabbage"), in the argot of schoolchildren, meant to have a temper tantrum. To have *les oreilles en feuilles de chou*, "ears like cabbage leaves," implies ears that stick out, which is saying a lot in Serge Gainsbourg's case.

"Why stupid as a cabbage?" wondered Marcel Proust. "Do you think cabbages are more stupid than anything else?" Something as stupid as a cabbage, we might respond, is within reach of the simplest among us, no doubt a trace of the negative connotation of rurality. Moreover, *berza*, Spanish for cabbage, has led to the term *berzas*, "idiot"; Italian *testa di cavolo* or English "cabbage-head" are further examples. When in the seventeenth century someone was sent to "plant his cabbages," it was a bad sign, indicating that he was being dispatched from the court to the countryside.

In horse-racing parlance, when a horse ends up "in the cabbages," it has gone off track; in other words, it's a loser.

*Ménager la chèvre et le chou*, "to handle the goat and the cabbage," means to avoid taking sides between two adversaries, through personal interest—and sometimes at one's own risk if one is to believe Mme de Sévigné: "He likes to handle the goat and the cabbages. He has handled the goat badly and will not even be able to eat the cabbage." Hence the pejorative *mi-chèvre, mi-chou*—"half-goat, half-cabbage."

---

*Quoted in Philippe Gillet, Le Goût et les Mots (1993), p. 125.
† I am a cabbage-headed man / Half vegetable, half guy.

The term of endearment *mon chou*, however, has nothing to do with the vegetable but is thought to derive from the verb *choyer*, "to pamper." In France, children are told that baby boys grow in cabbages and girls in roses. (It should be noted that in Alsace, the land of sauerkraut, all babies are brought by the stork.) For girls, the rose is the symbol of beauty but is also associated with the cult of Mary. For boys, should the cabbage simply be interpreted as a symbol of fertility? In some regions, cabbage soup is given to newlyweds to eat. Another explanation, implausible but better suited to questions children ask, is that a nice soft cabbage, sheltered by the layers of leaves, may serve as a better cradle than other vegetables. But here we may also see a trace of the Greek legend linked to the origin of the cabbage. The cabbage was believed to have been born of the tears of Lycurgus, king of Thrace, rendered insane by Rhea, goddess of the Earth. Taking his son for a vine sprout, the king was about to chop it down. His subjects overcame him and put him to death. Cabbages started to grow in the places where Lycurgus's tears had fallen.

There is still much more to be said about the dietary and medicinal qualities of the cabbage. It used to be called "the medicine of the poor." Cabbage juice was used in the Orne to treat stomach ulcers; cabbage leaves served as poultices. Cato the Elder declared that "it cured melancholy, it cured everything." Dried stalks of certain cabbages were fashioned into canes, most famously the one used by Charlie Chaplin in his movies, made out of a cabbage stalk from the island of Jersey.

But there has been a disturbing decline in varieties. In 1890 the Vilmorin-Andrieux family inventoried the vegetables cultivated in France. The list of cabbages, with a painstaking description of their origin, and of their mode of cultivation and reproduction, included 1,012 varieties.

Today the Vilmorin seed catalog lists only a few units of each type of cabbage. In 2007 Willemse Seeds listed thirty-three. Many seem to have completely disappeared, though a few have been preserved in a herbarium in the Museum of Natural History or in the collections formed since 1945 by the Institut national de recherche agronomique (INRA). These are used as reserves, from which different varieties are derived;

combined in a laboratory, the varieties become hybrids and are marketed under registered patents. The sale of seeds is strictly controlled in France. But still to be found in gardens are some ancient varieties, which are at risk of gradually disappearing, such as the perpetual cabbage, the only one not to flower or to reproduce year after year. Fortunately, trading is fully permitted, and the conservatories that practice the exchange of cuttings or plant seeds each year save dozens of local varieties.

Clearly, cabbage continues to feed us, as illustrated by a recent weight-loss fad, the cabbage soup diet, which consists in eating freeze-dried cabbage soup three times a day for one week. As Joseph Delteil wrote in *Les Cinq Sens*, "Cabbage soup is reputed to be a brain food, favorable to developing thought. It is good for people with rickets, for unwed mothers, and for rabbits."

# *Parsnip*

✤ ✤ ✤ ✤ ✤ ✤ ✤ ✤ ✤ ✤ ✤ ✤ ✤ ✤ ✤ ✤ ✤ ✤ ✤ ✤ ✤ ✤ ✤

This oft-forgotten root vegetable was for centuries one of the most commonly eaten. Along with cabbage and turnips, it formed a soup base.

Until the nineteenth century, the parsnip was inseparable from its cousin the carrot. But such is life; the couple separated, and the carrot, at least in France, was preferred. In Great Britain, the United States, and Portugal, however, the parsnip remained a popular vegetable.

Like the carrot, the parsnip belongs to what was formerly known as the Umbelliferae family, today called Apiaceae (from the name "celery," *apium* in Latin, the vegetable that gave the family its name). It is a large family, embracing some sixteen hundred species, including fennel, celery, parsley, chervil, coriander, caraway, cumin, green anise, angelica—all aromatic plants that flavor our cooking. They all share their origin in Europe or, at least, in the northern hemisphere, and seek shade and relatively cool temperatures. Parsnips even hold up to frost, which explains the appearance of certain varieties in regions such as Siberia and the Caucasus.

In their natural state, the carrot and the parsnip seem quite different. History, however, has increased the confusion between them, making their identification difficult. Both are very ancient, although we don't know exactly when they appeared. They must have been gathered in distant times, then domesticated—but when? They were known to the Greeks and the Romans but are hard to distinguish clearly through the texts of ancient authors, as their methods of recognizing plants weren't based on the same criteria as ours.

Several words are used to identify them, including *pastinaca*. But if for Athenaeus, a second-century scholar, they were indeed the same plant, the physician/botanist Galen identified the carrot separately and named it *Daucus pastinaca*.

The confusion between the carrot and the parsnip arose from the similarity in the shape of their roots and their whitish color—even though red carrots already existed in Syria. The parsnip has also been confused with chervil, though the two plants are similar only in their leaves. Chervil—originally from China or Japan, known to the Persians and then to the Greeks and the Romans—was enjoyed, according to Pliny the Elder, by the emperor Tiberius, who had discovered it in Germania. A single Latin word, *siser*, sometimes was used to designate chervil, carrots, and parsnips; terminology compounded the confusion caused by the different interpretations of various botanists. The sweet taste of the chervil root no doubt explains why German gives it the same generic name as the parsnip: *Zuckerwurzel* (sweet root). Its scientific name, *Sium sisarum*, derives from the Celtic *siw*, meaning "water."

———

We will limit ourselves here to *Pastinaca sativa*, which in Linnaeus's nomenclature designates the cultivated parsnip. According to the French dictionary Littré, the word *panais* (parsnip) comes from *panacem*, from the Greek word *panakeia*, which became "panacea," a remedy for everything. Le Robert, another French dictionary, shows a first appearance in 1080, in the form of *pasnaie*, a feminine word like its Latin model, *pastinaca*, which itself came from *pastinum*, a small hoe, an etymology that not everyone agrees with. Other authors believe it comes from *pastus*, "food," indicating both its popularity and its nutritional content. Indeed, parsnips do provide a lot of energy thanks to their glucides, fiber, potassium, and vitamin B.

The parsnip's common names in French poetically translate this confusion: *carotte d'hiver* (winter carrot), *carotte sauvage* (wild carrot), *grand chervis* (great chervil), or *pastenade blanche* (white carrot).

As one might imagine, these historical gropings have not prevented the parsnip's popularity: mentioned by Pliny the Elder, Columella, and Theodorus of Tarsus, planted in medieval gardens (*Le Mesnagier de Paris* grants it an important place in the chapter on cultivation), parsnips were widespread until the sixteenth century. Gilles de Gouberville, for example, had them planted in his vegetable garden in Cotentin; they were grown by Olivier de Serres in his estate of Pradel, and by La Quintinie, who distinguished them from carrots in the King's Vegetable Garden.

The vegetable's popularity can also be found in certain rituals. In the revolutionary calendar of Fabre d'Églantine, adopted in 1793, it has its own feast day, which was celebrated two days after Holy Carrot Day, on 9 vendémiaire, the first month of the calendar that began on September 22.

For some, the parsnip is even associated with two customs during Halloween, well before the introduction of the pumpkin during the adoption of the feast in the United States after the massive emigration caused by the 1848–49 famine in Ireland.

❖ ❖ ❖ ❖ ❖ ❖ ❖ ❖ ❖ ❖ ❖ ❖ ❖ ❖ ❖ ❖ ❖ ❖ ❖ ❖ ❖ ❖ ❖ ❖

### PUREE OF ROOT VEGETABLES

First, take some carrots, turnips, parsnips, and onions, and divide them, half to be grated with an iron grater: put the resulting pulp in water to be heated; after three or four boilings, pass everything through a muslin sieve, or a strong light cloth. Then, take the other half of the same roots, cut them lengthwise into thin pieces, and sauté in butter; then put them in the above-created liquid and cook. One can add to this bouillon, to give it more substance and make it more substantial, a spoonful of bean flour, peas, lentils, or other beans; or, better yet, rice. Keep in mind that vegetables intended for soup must always be grated first: in that state they provide all their nutritive qualities, and one needs fewer of them to obtain a greater quantity of nutritious food.

\* A medieval soup recipe provided by N. François in 1804.

❖ ❖ ❖ ❖ ❖ ❖ ❖ ❖ ❖ ❖ ❖ ❖ ❖ ❖ ❖ ❖ ❖ ❖ ❖ ❖ ❖ ❖ ❖ ❖

On the one hand, round parsnips that had been cleaned out could be used as lamps to celebrate the light during this Celtic celebration of the death and resurrection of holy fire; on the other, accompanying cabbage and fried onions, and surely well before the appearance of the potato in Europe, they were a component of calcanon, a traditional dish that was served that day in Britain and in Ireland. Symbolic objects, such as a ring or a thimble, were slipped in, playing an oracular role. Calcanon remains a typical Irish dish, but contemporary recipes, with a base of cabbage, potatoes, leeks, and onions, no longer seem to include parsnips.

These customs in any event translate the Anglo-Saxon taste for parsnips, pleasantly expressed by Samuel Beckett in his novel *First Love*: "I like parsnips because they taste like violets and violets because they smell like parsnips. Were there no parsnips on earth violets would leave me cold and if violets did not exist I would care as little for parsnips as I do for turnips, or radishes."*

In France, however, the parsnip lost much of its popularity through the centuries. It was gradually confined to a principal variety, the half-long Guernsey, while the carrot, for its part, gained ground.

Once again, the negative coefficient was the parsnip's association with peasant food or with fodder, even for pigs. Parsnips were fed to cows in the belief that they would increase the amount of milk produced—and hence to wet nurses too.

Such low status is reflected in French in the form of insults, such as *panais pourri* (rotten parsnip) on the Côte d'Or; or *panesennec* (imbecile) in Brittany; or in a phrase in one of Aristide Bruant's songs: *tu es un panais gelé* (you are a frozen parsnip). The word is also used in Louis-Ferdinand Céline's novel *Mort à credit* and elsewhere to refer to the penis.

———

The popularity ancient vegetables are now enjoying reflects the evolution of taste. The parsnip is making a comeback, if not on most of our tables, at least on those of chefs, on market stands, and even in supermarkets. As long as it is bought young and fresh, it is delicious. It is occasionally eaten as a dessert in England and Ireland, candied in honey. The Irish, who know everything there is to know about beer, make a fermented beverage from parsnips, which, without rivaling Guinness, does evoke its slightly sweet flavor. You can also make ersatz coffee out of parsnips by roasting them. Like the Jerusalem artichoke, the parsnip has unsuspected possibilities.

---

*Samuel Beckett, *First Love and Other Shorts* (*Grove Press*, 1994), p. 33.

# *Carrot*

✢ ✢ ✢ ✢ ✢ ✢ ✢ ✢ ✢ ✢ ✢ ✢ ✢ ✢ ✢ ✢ ✢ ✢ ✢ ✢ ✢ ✢ ✢ ✢

As we have seen, until around the sixteenth century, it was hard to tell a carrot from a parsnip. Ancient texts do little to clarify the distinction. Referring to the carrot, by the name of *carota*, the Roman cook Apicius describes it as a ligneous and whitish root, which he suggests should be served fried or in a salad. It also appears in bunches on the frescoes of Herculaneum. But when did carrots become carrot-colored? The history is truly unclear, since the details vary depending on the sources.

In the 1930s a team led by the Russian biologist Nikolai Vavilov, working in an extensive program to improve the plants grown in the service of Soviet agriculture, discovered types of spontaneous, cultivated, and hybrid carrots in Afghanistan and Kashmir. Their appearance differed from the wild carrots of our climates: their roots were fleshy, not divided, and their color ranged from purple and pink to yellowish orange. Additional prospecting in Anatolia produced less striking results along the same lines, which were due to an early migration of those populations toward the west, before or following the domestication of plants.

In the twelfth century, the agriculturalist Ibn al-Awwām reported in his Arabic treatise *Kitab al-fila-hah* that according to a fourth-century compilation made in Syria on Nabataean agriculture, farmers living in Palestine in the sixth century BC knew of "a red type and a yellow type, the red being finer, juicier, and more flavorful than the yellow." They created dishes out of carrots that were eaten with vinegar, brine, different vegetables, or grains. Other populations also used carrots to make bread, mixing them with wheat, rice, or millet flour.

Did nomadic merchants bring back those varieties of carrots from the East? Did the Romans know of them? If so, did they fail to mention them because they had already established and preferred the domestication of the white varieties?

The migration of these varieties to eastern Afghanistan happened rather late; they moved from Iran toward China in the thirteenth century, then to Japan in the seventeenth century. They would only later be crossed with Western varieties. Ibn al-Awwām points out the existence of red carrots in Spain, a fact confirmed by still current Judeo-Spanish, Majorcan, Portuguese, and Maghreban linguistic variations.

Hence, it is reasonable to hypothesize that between the tenth and the thirteenth centuries, Afghan carrots were transported by caravans from the East to Spain via North Africa. Since linguistic forms vary so widely, at least two other trajectories are possible: one through the center and west of Europe, the other through Slavic countries and Romania. At the end of the thirteenth century, Pietro de Crescenti mentions a root sold in bunches in Italy from which a fine red puree can be made. Presumably he is referring to carrots (though some think it might have been beets).

Around that time, the same type of carrot was found in Germany, in the Netherlands, and a little later in Britain. Renaissance botanists described the same characteristics as the Russian scientists in Afghanistan. Olivier de Serres claimed that "parsnips and carrots are practically the same except for their color: the one being red, and the other white," while insisting on the confusion of the names. Nicolas de Bonnefons in *Le Jardinier français* noted: "There are three different-colored carrots: white, yellow, and red. The yellow ones are the most delicate to cook."

Only La Quintinie, the gardener of Versailles, described the carrot as "a sort of root, some white, others yellow." He clearly distinguishes carrots from parsnips but doesn't mention any red ones, which might lead one to believe that at the royal table, yellow or white carrots were preferred over red ones, which "tainted the broth."

To sum up: in the seventeenth century, there were white, purple, and yellow carrots; and red carrots were less common in France than in the rest of Europe.

---

Flemish painting is partially responsible for introducing orange carrots as we know them today. Sixteenth-century still lifes by Pieter Aertsen and Joachim Beucklaer display various types of carrots surrounded by piles of cabbages and turnips. We see carrots from Anvers in Vincenzo Campi's painting *Christ in the House of Mary and Martha*, and yellowish-orange carrots in the foreground of the stall of Joachim Beuckelaer's *Market Woman with Fruit, Vegetables and Poultry*. But Joachim Wtewael's painting *Kitchen Scene* (1605) was the first to indicate the presence of these new orange carrots in the form of a few thin roots of an orange hue. Other painters of the same period—Van Rijck, Dou, and Battem—painted both types of carrots: some long and thin, others short and plump.

Flemish painting, then, plays an essential role as documentation, as the painters proved to be good witnesses of their times. Celebrating abundance and Holland's major role in market production, some paintings function as advertising posters.

Horticulturalist authors were clearly behind painters: it was only in 1721 that those orange carrots from Holland were mentioned elsewhere in Europe. They were first introduced into France around 1770.

We don't know the biochemical process that enabled the whitish types to turn into the orange type. Patient selection from cultivated yellow populations may have been the answer, since attempts using wild carrots always failed. It was during the nineteenth century that carrots really began to be differentiated. The white, purple, and yellow types became rarer, without disappearing entirely. In 1925, there were forty varieties before hybridization. In 1999, 93 percent of the varieties were obtained through hybridization. Today the carrot is one of the most widely cultivated vegetables in the world, especially in Japan and Africa, due to its high content of fiber, sugars, and pro-vitamin A, to which it has given its name: carotene. The carrot is believed to reinforce the immune system, and it is an antioxidant, excellent for the skin and vision.

Although the word "carrot" didn't appear in dictionaries until the sixteenth century, it soon figured in a large number of idioms and metaphors, thanks to its shape, color, and significance.

In 1723 the *carotte de tabac*, "tobacco carrot," appeared; it was a conical-shaped rolled leaf that we find in the *bureau de tabac* emblem. By extension, the carrot designates a cylindrical sample obtained by drilling into the earth.

The carrot's shape, like that of the parsnip, very early acquired an erotic connotation. In Saintonge, brides used to be given a large carrot with two goose eggs. In the Aisne, on May 1, a stick with a carrot and two potatoes was placed in front of houses where loose women lived. Many allusions to masturbation complete this tableau, showing an early interest in sex toys.

The carrot's orange color suggests above all the opprobrium to which redheads are prey. A famous example is Jules Renard's *Poil de Carotte* (1894). The first words of the novel are spoken by the mother, addressing her redheaded youngest child by her affectionate nickname for him: "Carrot Top, go lock up the chickens!" Renard was well aware that carrot-colored hair also had an evil connotation. Although carrot juice is recommended to pregnant women, a redheaded child may not turn out well. As for red-haired women, they were often considered witches, red being the color of the devil. Carrot-colored hair constituted a curse.

Although the carrot represents an incentive for the donkey (as opposed to the stick), and although eating carrots gives pink thighs or a good nature, the great majority of carrot idioms have a pejorative meaning.

Carrots are eaten during Lent, and then on Good Friday, with the red color evoking the blood of Christ, according to the "theory of signatures" inherited from antiquity. For the same reason, carrots are prescribed to women to hasten the onset of menstruation.

The carrot, like other roots (recall the medieval "great chain of Being"), was a vegetable of the poor until the Renaissance. "To live on carrots" meant to live poorly, meagerly; and "to shit carrots," according to the Furetière dictionary (1690), meant to be constipated. Such references to bodily functions, whether pertaining to digestion or to sexuality, occur only in colloquial language.

In a twist of meaning, *vivre de carottes*, literally "to live on carrots," designates someone who dupes people. A series of expressions relates to the notion of trickery. *Tirer une carotte*, "to pull a carrot," means to extract confessions through ruse or to ask for money under false pretenses; *tirer la carotte*, "to pull the carrot," means to play a short game and take very few risks (*Dictionnaire de l'Académie*, 1740); and *carotter le service*, "to carrot the service," means to shoot at the flank (1867). Going from the noun to the verb, we still note an insistence on the idea of swindling: *carotter*, "to carrot," means to steal, and a *carotteur* or a *carottier* is quite simply a crook.

Several hypotheses attempt to explain this pejorative sense. There may be a connection with the *carotte de tabac*, which long related to contraband activity. Or a connection with the tax that the governor of Savoy levied on every bunch of carrots, which could be paid in the form of two carrots, as Littré suggests. Let's conclude with the famous *boeufs-carottes*, the office of General Inspection of the Police, responsible for catching *carotteurs* in its ranks.

———

The carrot illustrates two currents in French cooking. On the one hand, shredded carrots incarnate the dietary cooking trends of the 1980s and '90s, which privileged health, the raw, cooking al dente rather than simmering or overcooking. On the other hand, stewed beef with carrots and other beef dishes are symbols of cooking that is bourgeois, rural, provincial, and nourishing, inherited from the nineteenth century.

Marcel Proust was happy to make *boeuf-mode* emblematic of a work of art in *Remembrance of Things Past*, as he expresses with incomparable grace in this letter to Céline Cottin, his cook. To enjoy hot. Or cold.

[ JULY 12, 1909 ]

*Céline,*

*I wish to compliment and thank you for the wonderful boeuf-mode. I hope I can be as successful as you in what I am going to do this evening, that my style will be as brilliant, as clear, and as solid as your aspic—that my ideas will be as delicious as your carrots, and as nourishing and fresh as your meat. While I labor to complete my work, I congratulate you on yours.*

*M.P.*

# *Pea*

✤ ✤ ✤ ✤ ✤ ✤ ✤ ✤ ✤ ✤ ✤ ✤ ✤ ✤ ✤ ✤ ✤ ✤ ✤ ✤ ✤ ✤ ✤ ✤

"Once upon a time there was a prince who wanted to marry a princess; but she had to be a real princess." So begins Hans Christian Andersen's fairy tale *The Princess and the Pea.*

The prince travels the world to find his princess, but in vain. So he returns to the castle. One night, during a terrible rainstorm, someone knocks on the castle door. It's a young girl, soaked to the skin. Yet she asserts that she is a true princess. "We shall see!" mutters the queen, cautiously. She goes into the guest room, removes all the bedding, and puts a pea on the bed. Then she piles twenty mattresses on top of it, followed by twenty featherbeds.

The next morning she asks the princess how she slept.

"Oh, very badly! I scarcely closed my eyes all night. Heaven only knows what was in the bed, but I was lying on something hard, so that I am black and blue all over my body. It's horrible!"

"The prince," adds the storyteller, "took her for his wife, for now he knew that he had a real princess; and the pea was put in the museum, where it may still be seen if no one has stolen it."

———

Peas have been around for a long time. The first traces of them, found in the Middle East, date from the Early Neolithic, between 7000 and 6000 BC. Their existence in their wild state is lost in the dark of time; seeds dating from the period 6000–5000 BC were already cultivated, which makes peas among the earliest cultivated plants in the history of humanity, as old as other legumes, lentils and chickpeas, wheat and barley. They were later found in Egypt, thus are native to the Fertile Crescent; they subsequently migrated toward Europe and Asia.

Greek and Latin authors—Theophrasus, Pliny the Elder, and Columella—mention them: the Greek *pisos* led to the Latin *pisum*, from which we derive the French *pois* and the English "pea." Scientific nomenclature adds *sativum*, which means "cultivated" (*Pisum sativum*).

In the Middle Ages, peas were a basic food, along with grains and beans. They were often dried, which allowed them to be preserved and to serve as a precious resource during times of famine. They were used in soups or, better yet, mashed with bacon.

*Qui a des pois et du pain d'orge*
*Du lard et du vin pour sa gorge*
*Qui a cinq sous et ne doit rien*
*Il se peut qu'il est bien.* *

Hot mashed peas were sold in the streets; they also constituted the "pittance" that was distributed to the poor at the doors of convents—a medieval form of the modern food pantry or soup kitchen. Different varieties already existed: green peas that were eaten dried or fresh, but also sugar peas, as revealed by Guillaume de Villeneuve in the thirteenth century, who heard merchants crying in the streets of Paris, "Fresh peas in their shells, right off the vine!" And Platina offered a recipe for peas and bacon that clearly required sugar peas.

✤ ✤ ✤ ✤ ✤ ✤ ✤ ✤ ✤ ✤ ✤ ✤ ✤ ✤ ✤ ✤ ✤ ✤ ✤ ✤ ✤ ✤ ✤

### PLATINA'S RECIPE FOR FASÉOLES

Boil the peas with their pods and salt only once, then remove them from the water; fry a few nice slices and bits of bacon that is neither too fatty nor too lean. Next, add the peas and fry everything together. Finally, add a little vinegar with some cooked or sweetened wine must, and this is how you prepare *faséoles*.

✤ ✤ ✤ ✤ ✤ ✤ ✤ ✤ ✤ ✤ ✤ ✤ ✤ ✤ ✤ ✤ ✤ ✤ ✤ ✤ ✤ ✤ ✤

---

*Whoever has peas and rye bread / Some bacon and wine for his throat / Whoever has five cents and owes nothing / Is most likely quite content.

Peas are mentioned in all the chronicles; they are present in fables and in tales like *Cinderella*, in which the unfortunate girl has to pick out the peas and lentils thrown into the ashes by her wicked stepmother. They were equally common in England and also in Holland, the country that produces the most popular varieties. Christopher Columbus took some peas with him across the Atlantic and had them planted in Santo Domingo. The painter Arcimboldo chose peas for the toothy smile of his *Summer* figure.

The term *petit pois*, peculiar to France and referring to the green pea as we know it today, is not mentioned until the seventeenth century, in *Histoire générale des plantes* by Jacques Dalechamps, a doctor from Lyons. But it had no scientific designation and didn't appear in seed catalogs of that time. The term was used by Parisian cooks and consumers starting in the second half of the seventeenth century, if we are to believe the article "Pois verts, Petits pois" in Diderot's *Encyclopédie*. It became current in the rest of France only at the end of the eighteenth century.

⸻

The pea's stellar career actually began on January 18, 1660. On that day, the head chef of the Countess of Soissons, a man named Audiger (and the author thirty years later of *La Maison réglée*), attempting to obtain a monopoly over the sale of fine liqueurs that he had learned to make in Italy, presented the king with a crate containing herbs, rosebuds, and peas in their shells. And this as early as January 18! The Count of Soissons, as a good courtier, started shelling them for the king.

A true frenzy took over the court, described thus by Mme de Maintenon: "The chapter of peas is still continuing, the impatience to eat them, the pleasure of having eaten them, and the joy of eating more are the three points that our princes have been making for four days"; and "there are ladies who, after having supped with the King, and supped well, find peas in their rooms to eat before going to bed, risking indigestion. It's a fad, it's a craze." They were eaten salted or, when served as a dessert, sugared. Doctors were quickly concerned about the possible effects of these delectable peas on the royal intestines.

Mme de Maintenon's description perfectly illustrates Flandrin's "distinction through taste" (see "A Matter of Taste," above). That which delighted and pleased the court was less the product itself than its early appearance, thus its rarity. To be able to eat peas in January—that was something that distinguished you from the bourgeois. This infatuation was satirized in plays and fables. The food historians Antoine Jacobsohn and Dominique Michel emphasize the specifically female taste for peas at the time, without being able to explain it. It became customary to include two pea dishes on a menu. There was even a recipe for *asperges en petis pois*, "asparagus peas," which first appeared in *L'Art de bien traiter* (L.S.R., 1674). The cook Antonin Carême offered it to Talleyrand, and it was borrowed by Gouffé (1807–1877) and then by Alexandre Dumas in his *Grand Dictionnaire de la cuisine*. This dish doesn't actually include any peas; it consists of asparagus cut up into small pieces, boiled in salted water, and then sautéed in butter with herbs or, in later recipes, with a béchamel sauce.

✣ ✣ ✣ ✣ ✣ ✣ ✣ ✣ ✣ ✣ ✣ ✣ ✣ ✣ ✣ ✣ ✣ ✣ ✣ ✣ ✣ ✣ ✣ ✣

#### ALEXANDRE DUMAS'S ASPERGES EN PETIT POIS

Use the smallest asparagus and cut everything that is tender into small pieces. Cook them briefly in salted water; drain them promptly in a sieve, and sauté them in a pan with butter, salt, pepper, and minced herbs, or put them in a pot, sprinkle flour and a little sugar over them, add a little broth or water, sauté them for a moment, and serve.

✣ ✣ ✣ ✣ ✣ ✣ ✣ ✣ ✣ ✣ ✣ ✣ ✣ ✣ ✣ ✣ ✣ ✣ ✣ ✣ ✣ ✣ ✣ ✣

Along with their early ripening, peas were judged by a second criterion, tenderness, which henceforth guided advances in their cultivation.

For Grimod de La Reynière, the pea was "the tenderest, the best, and the most delicate of vegetables," "the prince of side dishes." Flaubert, it appears, was a great fan of duck with peas. And Émile Zola makes that vegetable the highlight of the meal offered for Gervaise's feast day in *L'Assommoir*:

"Now, what sort of veg?"
"How about peas with bacon?" said Virginie. "I'd be happy with that and nothing else."

"Yes, yes, peas with bacon," the others all agreed, while Augustine, greatly excited, kept ramming the poker into the stove.*

Fresh peas became an unsurpassable dish in French bourgeois cooking. There was even a market at Les Halles specifically for peas.

It was not until the nineteenth century that canning—the process invented by Nicolas Appert to conserve food products—calmed the passion for fresh peas while revolutionizing the consumption of vegetables. The primary criterion then became delicacy, which guided selection. Production intensified, in particular in the region west of Paris, in the Marly area or on the hills bordering the Seine, as seen in Camille Pissarro's *Rameurs de pois* paintings (oarsmen hauling peas). Today peas are primarily eaten canned or frozen: in 2001, of the more than two kilograms of peas eaten per person in France, only 250 grams (about 9 percent) were fresh. Italy is now the primary European producer and consumer of fresh peas, or *piselli*.

Many varieties of peas are cultivated today, including peas developed for canning and for fodder.† Thousands of varieties exist throughout the world, the result of a patient work of selection. But they all share the characteristics of the legume family, or Fabaceae:

* flowers in the shape of butterflies
* the fruit in a shell
* the ability to absorb atmospheric nitrogen

The pea is an annual plant, autogamous, its characteristics vary depending on the variety; different varieties have white or purple flowers, long or short stems, smooth or wrinkled seeds, yellow or green pods, and so forth.

This relatively simple system enabled a revolution in our knowledge of the laws of heredity.

---

*Émile Zola, L'Assommoir, trans. Margaret Mauldon.

† *The French lists* pois fourragers, pois maraichers, pois mangetout, pois de casserie, pois de conserve, pois proteagineux.

Gregor Johann Mendel (1822–1884), born in Silesia to a peasant family and ordained a priest in 1848, devoted himself to teaching while pursuing his passion for science. His leisure time was occupied with physics, botany, zoology, paleontology, and, later, apiculture and meteorology. His work on hybridization continued for eight years, until 1870, when he became an abbot and had to abandon his research.

For his research on heredity, Mendel chose the pea because of its clear characteristics, its self-fertilization, and the ease of its hybridization. He was the first to see the key to solving what was considered an obstacle to hybridization: heterogeneous shapes that were obtained with the second generation of hybrids. He discovered that every characteristic of the first generation could have a heterogeneity, and that distinct characteristics—color and shape, for example—were transmitted separately and following precise rules.

We don't know whether Mendel envisioned the extension of his schema to animal hybridization. The importance of his work was unrecognized at the time. Not one of his colleagues seems to have sensed its merit, though the period was marked by evolutionism and the repercussions of Darwin's *On the Origin of the Species*. The fixed schema of Mendelian laws of heredity seemed to go against the current, insofar as he had been unable to distinguish between a characteristic itself and that which permitted its reproduction (the gene). It was only in 1900 that work carried out by other scientists would uncover "Mendel's laws," which today form the foundations of the science of heredity and genetics.

———

From prehistory to the genome, the pea has been a part of human history, not just as a food but also as a subject for wonder, the hero of a fairy tale or cautionary tale, a term of endearment ("My sweet pea," as Alain Souchon sings). We shell peas while daydreaming in the kitchen or chatting with the family. For Marcel Proust, peas "are lined up and numbered like green marbles in a game." For Philippe Delerm, in *We Could Almost Eat Outside*, "the shell of peas is not conceived to explain, but to follow the course, slightly against time."

———

Going back a bit in time, I'd like to end with a story.

Once upon a time in Normandy, a grandfather cultivated peas with blue flowers. Every summer, at the end of July, he harvested the dry seeds to plant the following year. One day, his granddaughter heard him complain that insects had destroyed his supply of blue-flowered peas for planting.

On Monday morning, he announced that he was going to get some at his seed merchant's stall at the market of Saint-Pierre-sur-Dives. At noon, he returned empty-handed: the seed seller no longer had that variety of blue-flowered peas.

For years the granddaughter saw her grandfather search in vain for his blue peas.

Later, after his death, she wanted to understand the reasons for the blue pea's disappearance. She saw that the variety no longer appeared in the national catalog of commercialized seeds. She took up the challenge to rediscover the blue pea.

She found it, in Belgium, where it is still cultivated.

Now grown up, that granddaughter devotes her life to saving ancient varieties that have disappeared from our catalogs and sometimes from our memories.

# *Tomato*

✤ ✤ ✤ ✤ ✤ ✤ ✤ ✤ ✤ ✤ ✤ ✤ ✤ ✤ ✤ ✤ ✤ ✤ ✤ ✤ ✤ ✤ ✤ ✤

It took the tomato two centuries to make its mark in Europe—and today it is one of the most widely consumed vegetables in the world, second only to the potato. The history of the tomato is full of excitement, has had major repercussions, and is certainly unfinished.

Originally the tomato grew in a wild state in the Andes, in Peru and Chile, in the form of small bunches, a bit like cherry tomatoes. It was discovered by the conquistadors in Mexico, where it was cultivated by the Aztecs, though we don't know exactly when it was first domesticated.

But we mustn't imagine that the Spanish conquerors were quick to sample that strange little fruit. Heroic, hard-headed, often bloody, generally fearless, those adventurers took risks of every kind *except* those involving food: the historian Madeleine Ferrières even suggests they were neophobic with respect to food. But for them it was a question of survival. So many unknown, even poisonous species or plants they didn't know how to prepare, such as manioc, which made people deathly sick because the Spanish neglected to peel it. And how could the soldiers of Christ imagine ingesting the same food as that eaten by those worrisome natives, some of them cannibals, even though the Valladolid debate suggested they were all brothers before God?

The entire culture of the period confirms this caution with regard to unusual food. Diets recommended consistency; any new food could be dangerous. Each plant possessed a specific nutritional and therapeutic value, established since antiquity; what was taught by botany was complemented by cooking. "It is not surprising that every day new illnesses, unknown in the past century, become manifest and are propagated from one land to another: indeed, we are adopting a new way of life, which we are importing from another world. For if we go looking for food in India, shouldn't we expect to be contaminated by it?" asserted Jean Bruyérin-Champier, a doctor from Lyon and the author of a history of food, *De re cibaria* (1560).

That doesn't mean, however, that the conquistadors weren't curious or eager to present samples of indigenous products to the rulers who subsidized them. Their expeditions provided an opportunity to bring back exotic products, strange plants, precious spices, fabulous animals, feathered savages, and so forth. Botanists undertook the description and naming of new species, classifying them by comparing them to known plants.

The conquistadors certainly brought the tomato back to Spain, but not instructions on how to prepare it, and for decades it was regarded with caution. From Spain it traveled to Naples, still under the Spanish crown, then up through Italy to Provence, the itinerary followed by the artichoke, as we have seen. But the tomato traveled slowly and was still not universally appreciated. For decades it was grown only as a curiosity.

The caution arose, no doubt, from the tomato's familial origins. It belongs to the Solanaceae family, which includes the petunia, tobacco, and the potato (the latter having suffered the same ostracism). Other members of that inhospitable family are poisonous plants such as deadly nightshade, stinkweed, henbane, or woody nightshade. It made sense to distrust the tomato, with its New World background. People were seriously afraid of being poisoned.

Even worse, the tomato resembled its relative, the mandrake. And the mandrake had a very bad reputation—black magic; a smell of death and sulfur; yellowish-orange berries that fall off when ripe; and then overnight, at total maturity, the plant vanishes. That was highly abnormal. The mandrake is also known as *mala terrestria*, "earth apple." It was whispered that it was born of the sperm of the hanged, under the gallows. Monstrous offspring, roots resembling legs, with the torso and sometimes the genitals of a human being. The mandrake belongs to witches. Recent studies have confirmed its hallucinogenic properties. "The mandrake is the preferred plant of witches for its psychedelic properties; it is the food of Sabbath gatherings," as Madeleine Ferrières is pleased to recall.

The agronomer Olivier de Serres advised against eating the tomato and was content to have it climb on the arbors of his pleasure garden. Unlike the squash and the Jerusalem artichoke, it was not on La Quintinie's list of vegetables and fruits planted in the King's Vegetable Garden. And unlike peppers or corn, it does not appear in Arcimboldo's portraits. Only a few later Spanish paintings such as Luis Eugenio Meléndez's *Still Life with Cucumbers and Tomatoes* (1772) pay homage to the tomato, proving that it was not yet common in markets or in kitchens.

Its names reflect the journey of the tomato. It all begins with *xi-tomatl*—the generic term in Nahuatl, the Aztec language, designating fruits or edible berries—to which the prefix *xi* is added, specifying the culinary preparation to be used. The word *tomatl* used alone describes the physalis, tart green berries that resemble unripe cherry tomatoes.

The *xi-tomatl* was given the name *tomato* when it came to Spain. Upon its arrival in Italy, the Italian botanist Matthioli renamed it the "apple of Peru," then *malum aureum*, or golden apple. (Originally, *malum* designated all fruits that were more or less round such as apples, oranges, lemons—whence the confusion regarding the famous "apple" held out by Eve to Adam, as the Hebrew word simply means "fruit.")

The "golden apple" suggests the tomato's color at the time. *Malum aureum* became *pomo d'oro*, or the Italian *pomodoro*.

When arriving in Occitania, or the southern part France, the golden apple became the love apple (*poumo d'amor*), either because it was believed to have aphrodisiac powers—as almost every new exotic vegetable did—or because it evoked mythological connotations such as famous golden apples from the garden of the Hesperides (identified most of the time as oranges), offered as a wedding gift by Gaia, the goddess of the Earth, to Hera. This poetic name remains in German (*Liebesapfel*) and in English ("love apples").

---

Used primarily as an ornamental plant or to repel mosquitoes, the tomato made a timid culinary appearance in Italy. Matthioli introduced tomatoes as "flattish and ribbed fruits that start green and become a golden yellow, which some eat fried in oil with salt and pepper like eggplant and mushrooms." A few brave souls at the beginning of the seventeenth century ate them in salads seasoned with verjuice. It was only in the course of that century that the tomato, under Spanish influence, became an element in Italian cuisine, which would give it its noble ranking: tomatoes were eaten in the Spanish style, only later used in sauces accompanying meat, and subsequently pasta. Tomatoes didn't appear in Neapolitan recipes until around 1692. The English author Lily Prior, in her delicious novel *La Cucina* (2001), gives us the traditional recipe for *strattu*, Sicilian tomato sauce.

✤ ✤ ✤ ✤ ✤ ✤ ✤ ✤ ✤ ✤ ✤ ✤ ✤ ✤ ✤ ✤ ✤ ✤ ✤ ✤ ✤ ✤ ✤ ✤ ✤

### RECIPE FOR *STRATTU*, SICILIAN TOMATO SAUCE

When all the tomatoes were sieved and there was no trace of skin or pips, we added some handfuls of salt and leaves of fragrant basil and poured the entire mixture onto the center of a large table we had set out in the sun behind the house.

Slowly, over a period of two days during which you must remember to stir frequently, the sun heats the mixture and evaporates the water, leaving a sumptuous, rich, dark tomato paste, which gives our pasta sauces their unique taste. ✳ LILY PRIOR, *La Cucina*, pp. 146–47

✤ ✤ ✤ ✤ ✤ ✤ ✤ ✤ ✤ ✤ ✤ ✤ ✤ ✤ ✤ ✤ ✤ ✤ ✤ ✤ ✤ ✤ ✤ ✤ ✤

Linnaeus gave the tomato the name *Solanum lycopersicum*, or "wolf-peach," and included it in his nomenclature in 1750. Its edible quality is apparent in the species name of the tomato as we know it today: *Lycopersicum esculentum*.

Nonetheless, in the eighteenth century, the tomato was still essentially unknown in northern Europe. Being delicate, it was difficult to transport and, more important, symbolized the cuisine of the south. The dishes that celebrate it today, ratatouille and pizza, for example, date only from the end of the nineteenth century.

It was the French Revolution that brought the tomato to Paris. According to legend, the Marseillais confederates in 1793, deprived of their fetish vegetable, reclaimed it wholeheartedly. It seems that tomatoes became fashionable in Paris thanks to restaurateurs from Marseille, especially the owners of Les Trois frères provençaux, on rue Sainte-Anne, and Le boeuf à la mode.

Grimod de La Reynière, in his *Almanach des gourmands*, inserted the tomato into the month of December, while noting that it stopped being any good at the beginning of that month. "Formerly rare and very expensive; they are now, at the beginning of this new century, being sold by great basketfuls in the Halles of Paris. Excellent in sauces to accompany meats, chopped up in soups with rice, they can even be eaten as side dishes," declares the gastronome, while providing a recipe, the first to appear in print, for stuffed tomatoes.

Clairvoyant as ever, he concludes: "We have no doubt that this pretty little fruit, given over to profound meditations of experts, will henceforth become the principle of a great number of learned and varied enjoyments."

Although Thomas Jefferson became interested in the tomato in 1781, it wasn't until the beginning of the nineteenth century that it took hold in the United States. Beginning its journey in Louisiana, it traveled north with the colonists, in the form of tomato seeds. Ketchup (from the Malayan *kêchap*, a fish sauce), far from being a recent invention, derives from a Louisiana recipe dating from that time.

It is thanks to the United States that the tomato is considered a vegetable. The authorities of the port of New York imposed a tax of 10 percent on all vegetables from the Antilles. Although importers argued that the tomato was a fruit, the court ruled firmly against them:

> Botanically speaking, tomatoes are the fruit of a vine, just as are cucumbers, squashes, beans, and peas. But in the common language of the people, whether sellers or consumers of provisions, all these are vegetables which are grown in kitchen gardens, and which, whether eaten cooked or raw, are, like potatoes, carrots, parsnips, turnips, beets, cauliflower, cabbage, celery, and lettuce, usually served at dinner in, with, or after the soup, fish, or meats which constitute the principal part of the repast, and not, like fruits generally, as dessert.[*]

---

[*] *Source: http://caselaw.lp.findlaw.com/scripts/getcase.pl?court=US&vol=149&invol=304.*

Although some present-day cooks such as Alain Passard are again regarding the tomato as a fruit, categorization of a food is generally determined by the way it is eaten.

———————

Through the forces of nature, the tomato's journey has caused genetic modifications. From the wild plant of the Andes to the first domestication in Mexico, from Mexico to the second domestication in Europe during the Renaissance, from the cultivated plant to the third domestication in the United States, numerous changes have been brought about through location, climate, and methods of cultivation.

Further modifications were linked to hybridization in the twentieth century: the first hybrid tomatoes appeared in the United States in 1920 and in Europe around 1950. According to Jean-Philippe Derenne, the tomato became "one of the battlefields of industrial genetics," with hybrids prevailing over fixed varieties.

The huge increase of these hybrids is impressive. In 1800 *Le Bon Jardinier* was happy to note laconically in a paragraph on the tomato that "there are large and small tomatoes." Six years later, seven varieties were listed in the Vilmorin-Andrieux catalog: "Large Red," "Early Red," Large Yellow," "Round Red," "Pear," "Small Yellow," "Cherry." We may note that these names concern color, size, shape, and early ripening. In 2000 the official French catalog listed 287 hybrid varieties and 30 fixed varieties. Two criteria of selection—resistance to cold and resistance to disease—were added to the ones mentioned above.

In France today, 90 percent of the tomatoes are produced in hothouses and pots. Specialists in these forms of cultivation use effective technical tools. Bees, for example, assure pollination, and ladybugs eat aphids. These very round, very smooth, very red, very healthy, impeccably calibrated tomatoes have only one problem: lack of flavor. The question of flavor (except, perhaps, for certain cherry tomatoes) remains crucial. Work is being undertaken to perfect a selection that would take into account all the ranges of flavor. We must have confidence in the engineers: they will one day succeed in creating a tomato that tastes like a tomato—the way they make croissants without butter that taste buttery.

The fragility of these hybrid varieties has required that zones where wild tomatoes grow in South America be classified as world biosphere reserves. Conservatory tomato gardens are being created, which may enable a return to the original varieties. Today around two thousand varieties of tomatoes exist in the world. Consisting of 94 percent water, thus very low in calories, tomatoes also contain lycopene, an antioxidant that fights the aging of cells and already exists in capsule form. Work is being undertaken to account for nutritional qualities in selection programs. So the tomato is promised a good future as a dietary supplement.

———

By way of conclusion, let us return to the symbolism of the tomato. Among certain ethnic groups such as the Bambara of Mali, tomato juice symbolizes blood, particularly menstrual blood, and hence fertility. Women offer tomatoes to the divinity, and couples eat a tomato before intercourse.

The Chilean novelist Isabel Allende ranks red, juicy tomatoes among aphrodisiacal vegetables. I recall being truly perplexed, as a child, by the image of a red tomato on a sign above a striptease club in Pigalle.

And Marcel Proust made comic use of that erotic connotation in *Sodom and Gomorrah*. Nissim Bernard, an old gentlemen who likes boys, avidly pursues tomato #1, a waiter at the Grand Hotel of Balbec with a head that irresistibly evokes the vegetable. Now the waiter has a twin brother, tomato #2, who shares his looks but not his tastes, whence a series of quid pro quos, in pure burlesque style. Nissim Bernard ends up getting sick on tomatoes.

But no one speaks of tomatoes more eloquently than Joseph Delteil, who was born in Corbières and became a vine grower in Massane, near Montpellier: "You are solar systems and the wombs of women, wombs of women and the brains of the earth," he writes in *Choléra*. In *La Cuisine paléolithique*, Delteil establishes a parallel between natural cuisine and raw art, summarizing, incidentally, my own concept of gastronomy: "the simple appetite between man and the world." One can't find a better example of this than his *Tomates à la Lucie*.

✤ ✤ ✤ ✤ ✤ ✤ ✤ ✤ ✤ ✤ ✤ ✤ ✤ ✤ ✤ ✤ ✤ ✤ ✤ ✤ ✤ ✤ ✤ ✤

## TOMATES À LA LUCIE

Take a few very round tomatoes, peel them, and put them in a pot over medium heat. Allow them to cook partially, but no more and no less; that is the secret: the heart of the tomato must remain raw, in its reddish skin. Its cheeks on fire and its heart cool. To complete, a topping of chopped parsley and garlic. Serve and pour all the juice over them. It reminds me of Scheherazade. ✳ Joseph Delteil, *La Cuisine paléolithique*

✤ ✤ ✤ ✤ ✤ ✤ ✤ ✤ ✤ ✤ ✤ ✤ ✤ ✤ ✤ ✤ ✤ ✤ ✤ ✤ ✤ ✤ ✤ ✤

THE

# *Bean*

"What is your country of origin? Did you come from central Asia with the broad bean and the pea? Did you belong to the collection of seeds that the first pioneers of cultivation brought us from their little gardens? Were you known in antiquity?" These questions, asked many years ago by the entomologist Jean-Henri Fabre (1823–1915), can now be answered today, in large part thanks to him.

For many years there was great confusion over the origin of the bean, of which no wild trace survives. Two theses have long been in opposition. The first sees an Asian or even Middle Eastern origin in these legumes, which have been consumed since earliest antiquity. The excavations of the nineteenth-century archaeologist Heinrich Schliemann in Asia Minor on the site believed to be the city of Troy revealed burned seeds of "common beans" mixed with chickpeas and beans from the marshes. The second thesis argues a much more recent introduction of the bean by the conquistadors, thus an American origin. But in that case, how could they have been eaten in Greece and Rome?

It would take an invasion of legume-devouring weevils for Jean-Henri Fabre to find the solution in 1901. These hitherto unknown weevils started to attack beans, while sparing European and Asian legumes. The common weevil, on the other hand, ignored beans, as observers had been noticing. Since for each plant there is usually a corresponding parasite, and since the weevils eating the beans proved to be of American origin, the plant seemed likely to have originated in America also.

So what were those legumes cited in ancient texts, from Theophrasus to Virgil, and still eaten in the Middle Ages?

Everything centered on the confusion between two plants: *Vigna* (*Dolichos*) and *Phaseolus vulgaris*. This confusion was fed by an unbelievable lexical imbroglio, due to the resemblance between these genera, which belong to the same family, the Fabaceae or legumes, differing only in their flowers. The Greek *phaseolos* gave the Latin *faselus*, which in French became *fasol, faziol, fayol*, then *fayot*. Until the seventeenth century, the term *fasiol* was used. Most often, the French words *pois* or *fève* are used interchangeably. We may remember that, similarly, the French term *légume* long designated all legumes: peas, favas, lentils, vetch, green beans, and so forth. As for the English "bean," the German *Bohne*, the Dutch *boon*, and the Swedish *böna*, they all come from a Germanic root *bauna*, "fava." Our modern need for scientific classification thus differentiates close species that, in the eyes of our ancestors, could be confused without harm. The distinction involved other factors as well: terms derived from the Germanic corresponded to a use of the seeds, whereas the derivations of *faselus*, in the Romance, Slavic, and Albanian languages, were connected with the sale of the pods in bundles (*faselus* probably came from *fascis*, "bundle").

Ever since antiquity, the *Dolichos*, a type of bean that came from Asia, was designated under various terms derived from the Greek *phaseolos*. This legume was commonly eaten until the Renaissance and survives in France in the Vendée and the Poitou-Charentes regions under the name of *mongette* (or *mogette* or *mojhette*). In Italy, Spain, Portugal, Brazil, and the Antilles, the *Dolichos* remains an important legume. Introduced into America by slave traders, it spread widely in the southern states, and it is found in all the tropical regions of Africa as well as Asia.

In *Tortilla Flat*, John Steinbeck remembers that for the poor in the Monterey region of California, "beans are a roof over your stomach. Beans are a warm cloak against economic cold."* Along with tortillas, beans made up the diet of Mexican families.

✢ ✢ ✢ ✢ ✢ ✢ ✢ ✢ ✢ ✢ ✢ ✢ ✢ ✢ ✢ ✢ ✢ ✢ ✢ ✢ ✢ ✢ ✢

### AN ANCIENT RECIPE FOR BEANS

In Tuscany there is a traditional way of preparing white beans (*fagioli*) that makes them particularly delicious. This process consists of filling

---

*John Steinbeck, Tortilla Flat (Penguin, 1977), p. 148.

a bottle two-thirds full of beans, or better, in a flask previously without its raffia dress, covering them with water, then hanging the flask above a steady fire of coal or wood and hot ashes, so that it hangs slightly at an angle. After eight to ten hours or more of very slow evaporation and cooking, the tender and velvety beans although whole and firm, can be eaten either all'uccelletto, that is, with a ragout sauce, or with olive oil and raw onions (which in my opinion brings out their very high quality). * Jean-François Revel, *Un festin en paroles*, p. 34

❖ ❖ ❖ ❖ ❖ ❖ ❖ ❖ ❖ ❖ ❖ ❖ ❖ ❖ ❖ ❖ ❖ ❖ ❖ ❖ ❖ ❖ ❖

Cultivated for six to eight thousand years from North to South America, beans were brought back to Europe by explorers and the conquistadors. Christopher Columbus discovered beans in Nuevitas (Cuba) that were very different from those known in Spain. Other explorers noticed them in Florida, Nicaragua, and even at the mouth of the Saint Lawrence River during Jacques Cartier's expedition. In South America, beans were cultivated along with gourds and corn. The explorers were not especially surprised by them, since they resembled a vegetable they were familiar with. They brought the plants to the monasteries of Seville, and in 1528 the monks sent them on to the pope. In Rome, the canon Piero Valeriano was delighted with their taste and with their ease of cultivation. At his insistence, Catherine de Médicis, a relative of Pope Leo X, agreed to take them on board the ship that transported her to Marseille, where she was to marry King Henry II in October 1533. The beans even formed part of her dowry. In France, then, thanks to the future queen of France, the exotic bean was grown in the vegetable gardens of Blois. Depending on the region, it was a kidney bean, a Roman bean, or a painted bean. For botanists or horticulturists like Olivier de Serres, it was the *fève de haricot* (fava bean), as opposed to the *fève de marais* (broad bean) for La Quintinie, who distinguished the two types in his King's Vegetable Garden.

"The little *fèves de haricots* or *callicots*, or *fèves rottes*, are of two different kinds: white and colored," writes Nicolas de Bonnefons in *Le Jardinier français* (1651). "You will notice the Reds above all others, greatly surpassing the Whites, although in Paris these are more favored."

But what about the term *haricot*? It appeared in César Oudin's Franco-Spanish lexicon in 1640 and came out of a new confusion between an

Old French word, *haricoter* or *halicoter*, which means to cut into small pieces (*haricot de mouton* was a mutton stew), and the Aztec term *ayacotl*, which designated the plant. The French poet José Maria de Heredia (1842–1905), of Cuban origin, was the first to find a trace of this etymology in a sixteenth-century book on natural history. But according to the lexicologist Alan Rey, the vague assonance with the Aztec term was probably pure coincidence; in 1620 terms such as *fève d'aricot* or *febve d'haricot* could be found, which indicated that legume was an accompaniment to the *haricot de mouton*. By slippage of meaning, *haricot* ended up designating the bean itself. Littré gives the same explanation.

Nicolas de Bonnefons suggests another hypothesis, in a recipe for *fèves rottes* (what in English might be called string beans): "They are eaten as *haricots à la nouveauté*, that is, with the shell; if they have strings at the ends of the two sides, you pull them, after which you parboil them, then fricassee and season them the same as peas without the shell, and also use cream to thicken the sauce."

———

Dried beans, which are easily preserved, used to be the last vegetables eaten at sea, after fresh supplies had been consumed. Because of their reputation for causing flatulence, their French name, *haricots secs*, has been used as an epithet for petty officers who were bullies and for sycophantic students.

At the end of the sixteenth century, it was customary in certain regions to serve beans during funerals, associating them with death. Should we see in this the survival of the symbolism linked to the bean? For Pliny the Elder, beans contained the souls of the dead. They were associated with cults of the dead, and with Saturnalia, elements of which survive in festivals associated with Christmas and Epiphany, as well as with the offerings of first fruits to God, so that the living might be protected by the dead. The same symbolism is found in Japanese tradition. When seared, beans were supposed to possess qualities of protection and exorcism. It was customary to scatter them around the house on February 3, in order to chase away demons.

There is religious symbolism in the legend of the Holy Spirit bean, also called the nun's bean or even the Communion bean, depending on the region. Here is the Orne version: In 1793, during the French Revolution, a priest on the run entrusts a monstrance to an old peasant woman and asks her to bury it. Imagine her surprise the following year when she harvests beans that she has planted on the same spot, and sees that they bear the design of the monstrance, showing the white circle of the eucharistic bread, surrounded by rays of light. In other versions, the beans show a mark in the shape of a bird with spread wings, representing the Holy Spirit. Though apparently miraculous, the image merely resulted from a natural hybridization from an earlier generation. On Good Friday, the purple seeds with white tips are dipped into holy water, to be planted only on the feast of the Holy Sacrament, in June. There is an analogous Tupi legend, in which a little girl, Mani, is buried in her grandfather's casket. A plant grows out of it, with a root similar to the body of the child. The manioc is born.

Whether green or dried, the bean evokes sexual imagery. Jean-Luc Hennig cites a quatrain by the Baroque poet Théophile de Viau:

> Je voudrais, belle brunette,
> Voyant votre sein rondelet,
> Jouer dessus de l'épinette
> Et au-dessous du flageolet.*

Isabel Allende recalls that "to Teutons and Romans the bean was a stimulant and its flower symbolized sexual pleasure. Bean soup had such a high reputation for being erotic that in the seventeenth century beans were banned from the Convent of Saint Jerome in order to prevent inopportune excitation."[†]

The popularity of a legume that became a staple in the lives of many is reflected in such idioms as *gagner des haricots*, "to earn peanuts" (literally "beans"), or *être maigre comme un haricot vert*, "to be as skinny as a green bean."

Yet it was not until the second half of the seventeenth century that the bean began to be written about. Until then only lentils, peas, and fava beans were mentioned in the local archives of Lower Normandy. A

---

*I would like, beautiful brunette, / Seeing your round breast, / To play the spinet on it / And beneath it the flageolet.

[†] Isabel Allende, Aphrodite: A Memoir of the Senses (Harper Perennial, 1999), p. 190.

century later, Combles notes in his work *L'École du jardin potager* (1752): "This plant is universally known and it is consumed widely in all countries. This vegetable is used so frequently in cooking that one can call it a major staple in the household." In 1733, as witnessed in the report of the intendant of the Généralité of Caen, beans appeared alongside other legumes in cultivation.

Was it simply a matter of vocabulary? As often happens, there was a gap between popular practice and subsequent scholarly discourse. Unlike tomatoes and potatoes, beans were adopted quickly, thanks no doubt to their kinship to peas, fava beans, and *Dolichos*, as well as to the ease with which they could be grown. They looked familiar to Europeans in spite of their exotic origins. They appeared as a crop vegetable as well as in family vegetable gardens. We also owe to Combles the first description of their morphological variability, as we still know it today:

* dried bean or fava bean (seed)
* shell bean (or tender seed)
* green or white bean, from which the string must be removed, or skinless (which can be eaten whole), dwarf or with runners

The bean belongs to the genus *Phaseolus*, which includes fifty-six species, four of which are cultivated:

* the common bean (*Phaseolus vulgaris*)
* the runner bean (*Phaseolus coccineus*)
* the lima bean (*Phaseolus lunatus*)
* the tepary bean (*Phaseolus acutifolius*)

To which we should add the mung bean (*Phaseolus mungo*), native to India and cultivated since time immemorial in China. The soybean, *Glycine max*, is a strong allergen. Whether it is offered in the form of oil, milk, seeds, flour, sauce, or paste, half of its production is genetically modified.

But if green beans were already known in the seventeenth century, a taste for them developed only later. Native Americans ate only the seeds, which became a primary foodstuff for the American colonists, who carried them in their wagons. In the nineteenth century, Henry

David Thoreau devoted an entire chapter of *Walden* to the "beanfield," which he made into a symbol of his agricultural labor.* "What shall I learn of beans, or beans from me?" he wondered. He learned the value of manual labor, of effort, the power of nature. But he also discovered, through a simple field of beans, his own roots and those of his land: "An extinct nation had anciently dwelt here and planted corn and beans ere white men came to clear the land." In hoeing his plantings, he disturbs "the ashes of unchronicled nations who in primeval years lived under these heavens." Thoreau thus has the intimate experience of nature and history.

The European bourgeois, who loved early fruits and vegetables, was responsible for turning the green bean into a sought-after vegetable, increasingly tender, thinner, a symbol of thinness, freshness, and refinement.

Marcel Proust became the herald of this fashion in *La Prisonnière*, when the gourmand Albertine delights in the "cries" of the street merchants: "And to think that we must still wait two months to hear: 'Green beans and tender beans, here are your green beans.' Well may they be called tender beans! You know that I want them very thin, very thin, dripping in vinegar; they look too good to eat, fresh as a rose."

Green beans, fresh as dew, thousands of tons of them, are now sent to European countries from Africa (Morocco, Egypt, Kenya, Senegal, and Tanzania), enabling us to eat them all year long, but not without harm both to African food crops and water resources and to production in Europe, where labor is more costly.

Homage should be paid to the inventor of the process that would make beans the champions of canning: Nicolas Appert. The son of an innkeeper from Châlon-sur-Marne, Appert had responded to an appeal from the army of the Directory during the Italian campaign: 12,000 francs to whoever could discover a way to conserve food. He had been a brewer, an *officier de la bouche*† of the Duke of Deux-Ponts, then a candy

---

*The quotations from* Walden *are drawn from the 1999 Oxford University Press edition, pp. 140, 141, 143.*

†*An* officier de la bouche *was not a chef; he was responsible for everything from his employer's bedtime snacks to catering immense banquets, including all the entertainments that seventeenth-century aristocrats loved: waterworks, fireworks, masques, stage illusions.*

maker on the rue des Lombards in Paris, before becoming mayor of
Ivry-sur-Seine. He bought some land in Massy, filled bottles with milk,
with beans, and with peas, and boiled them for varying amounts of
time in order to test the foods for preservation and flavor. He had discovered sterilization, or "canning." The British borrowed the process
and improved it by inventing the metal can.

Named "benefactor of humanity" in 1822 by the Society for the En-
couragement of National Industry, Nicolas Appert died a pauper in 1841.

But well after the invention of high-pressure canning, beans were
preserved in the countryside the same way as pickles, with salt and vinegar, in stoneware pots.

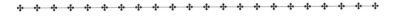

### BEANS WITH SALT AND VINEGAR

Place newly picked beans, carefully aligned, in a stoneware pot. Cover
with a layer of coarse salt, and continue alternating layers of beans
with layers of salt. Then add a liter of water and one-third of a liter
of vinegar per kilo of salt. Press down and cover with a hermetic seal.
The beans will keep all winter and will stay very crisp.

Afterward, simply rinse and cook as if fresh.

As Alain Rey points out, the bean illustrates "the borrowings and cross-
ings between cultural habits and traveling vegetation." Cassoulet from
Toulouse, *mogette* from the Vendée, Tuscan *fagioli*, Mexican chili con
carne, Brazilian *tutu* or *feijuada*, American baked beans, Caucasian *lobio*,
Indian *dhosas*, Kenyan *irio*: the list goes on. All around the world, beans
are eaten, red, green, black, yellow, mottled—almost as varied as the
populations of the Earth.

# *Pumpkin*

✤ ✤ ✤ ✤ ✤ ✤ ✤ ✤ ✤ ✤ ✤ ✤ ✤ ✤ ✤ ✤ ✤ ✤ ✤ ✤ ✤ ✤ ✤ ✤ ✤ ✤ ✤

New York, the Union Street Market. It's the middle of October. Basking in the light of the Indian summer are hundreds of pumpkins and other gourds of all sizes and shapes, in all shades of color from bright orange to pale yellow—shining like suns on the pavement. It is truly a wonder to behold.

* A squash or a pumpkin?*
* A marrow squash, or a zucchini?
* A gourd or a calabash?

It doesn't really matter, when you love them all . . .

———

Thanks to the work of Charles Naudin, a botanist at the Paris Museum of Natural History, we are fortunate today to be able to distinguish the various species of the genus *Cucurbita*. Kudos, then, to that lover of squashes, who in 1860 sorted out the thousands of members of the Cucurbitaceae family, which, from cucumbers to melons, pickles to watermelons, marrow squashes to pumpkins, have been spreading

---

*Pumpkin, which for most Americans refers to the large orange fruit we carve for Halloween and eat at Thanksgiving, is the common name for the genus* Cucurbita *of the family* Cucurbitaceae *(gourd family), a group that includes the pumpkins and squashes; the names may be used interchangeably and without botanical distinction. The French* potiron *and* citrouille *are both translated as "pumpkin," but* potiron *is "pumpkin" in the botanical sense, i.e., interchangeable with "squash." * Citrouille *seems to be used more specifically to refer to the American "pumpkin."*

their stems and tendrils over the ground in our gardens since time immemorial.

There are some dozen species of *Cucurbita* alone. The giant of the vegetable garden, the pumpkin, well deserves the name *Cucurbita maxima*. Some varieties, such as the Atlantic giant, can weigh hundreds of pounds. But they are grown only for show, or for photo ops in the local newspaper. There are numerous varieties of pumpkins and squashes, from the turban squash (named for its crown), to the delicious Hokkaido squash, a Japanese variety of Hubbard squash with wonderfully sweet flesh. I used to think that its French name, *potimarron*, came from the way it tastes, a mixture of pumpkin and chestnut, but apparently it came from the "chestnut squash," a Brazilian variety introduced by the Portuguese.

Pumpkins are not always orange. They can be yellow, blue-gray, dark green, bright red. They can be ribbed, bumpy, smooth, round, shaped like a top or a pear. The pumpkin has been prized above all for its aesthetic qualities: gardeners at the Château de Villandry or in the King's Vegetable Garden at Versailles would carve geometric designs on young pumpkins as they were growing. We can understand why pumpkins are so often seen in Flemish paintings or in the work of the Arcimboldo, who used a pumpkin to coif his "Autumn" figure—an "artist's pumpkin," as Aïté Besson noted, though it's difficult to determine the variety.

A second species of the genus *Cucurbita* is the *C. pepo*, the pumpkin most familiar to Americans. Although Jean-Baptiste de La Quintinie described the pumpkin a flat, yellow vegetable, for us it evokes Cinderella's magic coach or the grinning faces carved on Halloween pumpkins. But the *Cucurbita pepo* (baked by the sun) has many other faces: in addition to the American pumpkin, there are green and white squashes in the shape of a ribbed beret. Some squashes are long (like zucchini), spherical, or yellow. Spaghetti squashes have filaments that look like pasta when cooked; acorn squashes range from black to green and ivory; some varieties are grown for their seeds, such as the Lady Godiva, from which pumpkin oil is extracted, very popular in Eastern Europe.

A third species is *Cucurbita moschata*, or Moschata squashes, two varieties of which I find especially delicious: the sweet berry and the butternut, with pinkish-beige skin and bright orange flesh—as delicious with white meat or game as in a cinnamon-spiced pie.

Members of these species include the Siamese squash, its flesh similar to that of a watermelon; the Mexican squash, a pear-shaped cream-and-green variety; the luffa or towel squash (*Luffa aegyptiaca* L.), which absorbs liquid as it cooks; nor must we forget the calabashes and the gourds. But that's another story . . .

---

Like beans, squashes have a dual origin. Known since antiquity, they are mentioned by Pliny, Columelle, and other favorite authors or food lovers. Apicius offers as many as thirteen recipes for Cucurbitaceae, most often boiled, then fried, recooked into a sauce, and reduced into a puree. The skins of these squashes harden when drying out. Once emptied, they have multiple uses as receptacles, utensils, masks, birdcages, musical instruments, or even lifesavers for children in ancient Rome. They serve as drinking vessels for the pilgrims of Saint-Jacques-de-Compostelle, and also for the *maté* drinkers in South America, the term *maté* referring to both the vessel and the beverage. These calabashes or gourds are members of the genus *Lagenaria*, akin to but different from the *Cucurbita*. Experiments have shown that their seeds can still germinate after the fruit has been floating in water for a year, which doubtless explains why these squashes are so widespread. Originating in Africa, spreading very early into the Americas and into Asia, they are among the most ancient plants cultivated by humans (traces dating from 11,000–13,000 BC have been found in Peru). They have been part of the daily life of those peoples for thousands of years, generating myths and tales, especially in African traditions.

*Cucurbita*, on the other hand, originated in America and—like tomatoes, beans, corn, and peppers—were taken back to Europe by the conquistadors, who were amazed at their variety and beauty. On December 3, 1492, on a mountaintop in Cuba, Christopher Columbus discovered "a feast for the eyes." And Jacques Cartier expressed his admiration while exploring the Saint Lawrence River: "They have many large melons and cucumbers, squashes, peas, and beans of every color, not like ours." Squashes may first have been cultivated for their seeds, which are rich in oil. Traces have been found in Mexico and Peru dating from five thousand to seven thousand years ago. They spread via the Saint Lawrence to South America, where their seeds (*pepitas*) became a currency of exchange, and gradually people began to eat their flowers, their seeds (ground up), and their flesh. Transported by the conquerors of the New World, they reached Europe; thanks to the Portuguese, they were taken to Angola and Mozambique, then to India, Indonesia, and China, before returning to Europe by way of the Ottoman Empire and the Balkans. In a half century, the Cucurbitaceae had traveled the world, becoming acclimated wherever they went.

The diversity of their names is equaled only by the confusion it causes: Who can say what exactly the botanists of the seventeenth century had in mind when they distinguished the "pumpkin" from the "squash"? One French word for pumpkin is *citrouille*, from *citron* (lemon); another word, *potiron*, suggests various etymologies. Does it come from Syriac *pāturta* (mushroom)? From Latin *posterior*? From Old French *boterel* (toad)? Or from the adjective *pot* (swollen)? I rather like the last hypothesis, which brings to mind Jean-Pierre Coffe's favorite recipe. Master chef, TV personality, and defender of classic French food, Coffe makes a soup cooked right in the pumpkin.

✤ ✤ ✤ ✤ ✤ ✤ ✤ ✤ ✤ ✤ ✤ ✤ ✤ ✤ ✤ ✤ ✤ ✤ ✤ ✤ ✤ ✤ ✤ ✤ ✤

### JEAN-PIERRE COFFE'S PUMPKIN SOUP

Cut a lid off the top of the pumpkin. Add slices of toasted bread, mushrooms, grated cheese, crème fraîche, and nutmeg. Replace the lid and wrap the pumpkin in aluminum foil. Bake for two or three hours. Stir the cooked mixture, and serve.

You'll see—it's more than just edible!

✤ ✤ ✤ ✤ ✤ ✤ ✤ ✤ ✤ ✤ ✤ ✤ ✤ ✤ ✤ ✤ ✤ ✤ ✤ ✤ ✤ ✤ ✤ ✤ ✤

Following a brief infatuation, however, French chefs began to shun pumpkins. "They have no other use in cooking than to make cream soup," muttered Menon in *La Cuisine bourgeoise* in 1774. Since then, a lot of water has flowed over our Cucurbitaceae, which today enjoy a popularity due as much to their gustatory qualities as to their fantastic, polymorphic appearance. Pumpkins and other squashes attract thousands of visitors to garden shows every autumn.

In European idioms, "squash" and "gourd" are unflattering terms; they imply an empty skull, rather than a brilliant mind—one may note in passing that they are applied most often to females; the French term *gourde* describes not only a foolish woman but also a clumsy, awkward one. The calabash is a metaphor for the king's fool. But the opposite is true in other civilizations: in parts of Africa, for example, squash seeds imply intelligence and fertility. In the Far East, those prolific seeds symbolize abundance and fecundity. As sources of life and regeneration, they are inscribed in the Taoist tradition as a principle of immortality.

This leads us to the traditions of Thanksgiving and Halloween. In 1621 the Pilgrims celebrated their gratitude to Providence and to the Indians who welcomed them by adding the pumpkin to their menu, as well as corn and beans, which the Indians had taught them how to grow. The traditional Thanksgiving meal on the fourth Thursday in November usually includes stuffed turkey and pumpkin pie. In the original recipe, the pumpkin was emptied of its seeds, filled with apples, sugar, and milk, and baked in the oven.

Was it by way of the Thanksgiving tradition that the pumpkin made an entrance into the Celtic festival of Halloween? We may remember the role the parsnip traditionally played here (and probably also the turnip). The feast of Samhain, believed to have marked the beginning of the Celtic New Year, occurred at the end of October, when the herds came in from the pastures and the winter season began. Samhain celebrated the passage from light to darkness, from warmth to cold, from life to death. To enable a harmonious passage, the dead were invited to warm themselves among the living; to guide them, lanterns were lit and doors left open. But to prevent the intrusion of evil spirits following behind them, large fires were lit, which gave rise to gatherings and ritual games. A Dionysian feast that was deeply rooted in Celtic territory, Samhain resisted Christian attempts to replace it with All Saints' Day. Now known as Halloween or All Hallows' Eve (the day before All Saints' Day), the pagan celebration survived, blending the festive aspects with the macabre: skeletons, witches, and demons. For centuries, children have played a major role, perhaps because they are the surest antithesis of death. To scare old people, children would walk around at night carrying lanterns made out of turnips, parsnips, or (as in Brittany) beets, on which human faces were carved—jack-o'-lanterns. In the nineteenth century, pumpkins, which were harvested at that time of year, replaced other vegetables, first in America and then in Europe. The jack-o'-lantern gave its name to a variety of pumpkin particularly well suited for that use.

And so the pumpkin is intimately linked to rituals of passage: from the Old World to the New, from shadow to light, from life to death. It was quite natural for Charles Perrault to choose a pumpkin as a means for transporting Cinderella from the ashes in the fireplace to the ball, transforming the mistreated girl into an adored princess. Fairies are never wrong. Cinderella's fairy godmother is a queen of the night. She needs only to touch the pumpkin with her magic wand to turn it into a carriage and Cinderella into a princess. The pumpkin, an ordinary vegetable from the kitchen garden, brings the realm of reality into that of the marvelous. I don't know whether the tale is of Celtic origin. But some have seen in this story, which came out of an oral tradition, the survival of rituals that might relate to Samhain: Cinderella incarnates the New Year; first "betrothed to the ashes," humiliated by her wicked stepmother (who represents the year that has just passed), she finds a shoe that fits, thanks to a magical helper. From this perspective, the pumpkin fulfills its role as an instrument of passage. (By way of an aside: in the Germanic version collected by the Brothers Grimm, it is birds that transport Cinderella to the ball. Neither fairy godmother nor pumpkin appears.)

# *Chili Pepper*

✤ ✤ ✤ ✤ ✤ ✤ ✤ ✤ ✤ ✤ ✤ ✤ ✤ ✤ ✤ ✤ ✤ ✤ ✤ ✤ ✤ ✤ ✤ ✤

In the beginning, there were four brothers. The eldest, Manco Capac, was the first Incan. The second, Ajar Cachi (*cachi* means "salt"), incarnated knowledge. The third, Ajar Uchu (*uchu* means "chili pepper"), joy and beauty; the fourth, Ajar Sauca, happiness.

Thus begins the creation myth as told by Garcilaso de la Vega, the bastard son of the conquistador Sebastián Garcilaso de la Vega and the Incan princess Isabel Chimpu Ocllo. The most recent research confirms that chilies serve as a stimulant, that they "pepper things up," as Uchu might have said. Homeopaths use them to treat homesickness; Chinese pharmacopoeias, to treat depression. So we have good reason to conclude this book with a great traveler that has proved its worth in the four corners of the globe.

———

"Better than our pepper": with these historic words, Christopher Columbus greeted the appearance of the plant in Cuba, according to Pierre Martyr, the historian who accompanied him. Having gone in search of a new route for the spice trade, Columbus could hardly ignore this one. Like the bean or the squash, the *aji* had already been known for millennia by hunter-gatherers in Mexico and Peru. It would take an even more remarkable path in its world conquest. Like the squash, it owed its success largely to the Portuguese. It had the same point of departure as the squash, the ports of Bahia and Pernambuco in Brazil; the same Atlantic crossing, toward the Gulf of Guinea and the trading posts of Mozambique; the same route toward Goa and Calicut in India.

At the same time, however, the Spanish were taking a route in the opposite direction, and the chili pepper reached southern China and

Manila by way of the Pacific, starting out from Acapulco and Lima. When some Portuguese trading posts fell into the hands of the Turks, the chili pepper spread into the Ottoman Empire before reaching Moravia and Hungary, thanks to Spanish Franciscan monks. Persian, Malan, Hindu, and Arab merchants then spread it throughout all their zones of influence and commerce. Starting in Spain and the Maghreb, Western Europe was smitten with the chili pepper.

Far from being treated with caution like the tomato and the potato, the chili pepper was adopted everywhere. Was it because it contained no toxic elements? Because it was easy to grow and to preserve? Because it served as an inexpensive substitute for the condiment? Indian pepper, Calicut pepper, Guinea pepper, Cayenne pepper, Brazilian pepper, even poor man's pepper were some of its names. Chilies also benefited from their amazing capacity for adaptation, so that cultivated varieties could be re-naturalized, in India, for example. Try telling a Korean (Koreans are the current biggest consumers), a Chinese (China is the largest producer), or an African person that the chili pepper was not always a part of their food supply! Mexican chili, South American *aji*, Senegalais *pilipili*, Hungarian *paprika*, West Indian *zozios*, English "pepper," Russian *peretz*, German *Pfeffer*: the chili pepper speaks every language. And it is put into all kinds of sauces: *salsa mexicana verde*, Brazilian *carioca*, North African *harissa*, Louisianan *tabasco*, Caribbean *rougaille*, Indian *cari*, Indonesian *sambal*.

---

Not all languages, however, distinguish the chili pepper from the bell pepper. In Britain and North America, for example, "red pepper" is the generic term; "chili pepper" refers to the hot variety, and "bell pepper" to the mild. Both belong to the genus *Capiscum*, and their names reflect usage rather than true botanical differentiation. Certain species or varieties have been developed for their "incendiary" power, due to capsaicin, while others have become fleshier and sweeter.

✤ ✤ ✤ ✤ ✤ ✤ ✤ ✤ ✤ ✤ ✤ ✤ ✤ ✤ ✤ ✤ ✤ ✤ ✤ ✤ ✤ ✤ ✤

## VEGETABLE TAGINE

¼ cup olive oil
3 lbs. lamb or beef, cubed
2 medium onions, sliced
5 sprigs parsley, finely chopped
5 cloves garlic, finely chopped
1 teaspoon salt
½ teaspoon paprika
½ teaspoon pepper
¼ teaspoon saffron
½ teaspoon ground ginger
1 cup water
2 lbs. carrots, sliced
1 lb. white turnips, chopped
5 medium potatoes, peeled and sliced
[3 small chili peppers, chopped]

· Put oil in pan; add meat, onion, parsley, garlic, salt, paprika, pepper, saffron, and ginger. Sauté for 5 minutes. Add water, bring to a
boil, cover, and simmer for 30 minutes. Add vegetables and simmer
for another 30 minutes. You should have a small amount of sauce
left; if not, add ½ cup of water. Serve with couscous or rice. Serves
8. (*http://recipes.wuzzle.org/index.php/75/2310; the bracketed in-
gredient has been substituted for "1 medium green pepper" in the
original.*—TRANS.)

✤ ✤ ✤ ✤ ✤ ✤ ✤ ✤ ✤ ✤ ✤ ✤ ✤ ✤ ✤ ✤ ✤ ✤ ✤ ✤ ✤ ✤ ✤

*Piment*, the most general French term, comes from the Latin *pigmentum*:
it reminds us of the color properties of the pepper, which is increasingly
used today. The Spanish term *pimiento* encompasses all chili peppers,
whereas *pimienta* refers to the condiment pepper.

The genus *Capiscum* includes twenty-two wild and five domesticated species. Among the latter is *Capiscum annuum*, to which all the varieties grown in our gardens belong. Other species may still be discovered in the Amazonian region.

The Espelette pepper has become very fashionable on Paris dinner tables in the last few years. Traditionally cultivated by women, harvested in September in the little Basque village from which it derives its name, it is put out to dry in garlands on the fronts of houses and blessed during a festival at the end of October. We are less familiar with the red pepper from Bresse, once cultivated with beans, potatoes, corn, and squash and rediscovered some twenty years ago. Unlike the Espelette growers, however, the women of Bresse, who otherwise ruled over vegetable gardens, were forbidden to touch the peppers.

———

No one knows with certainty why the hottest peppers were most popular in poor, hot countries. Perhaps because they cost little but allowed people to add a bit of color, taste, and spice to simple foods: they are remarkable taste enhancers.

Spiciness, moreover, creates an attachment and a form of dependency. The capsaicin releases a series of painful reactions in our mouths that correspond to mechanisms of protection against poison. The brain then secretes endorphins to ease the pain. It is the mixture of suffering and pleasure that becomes addictive. Those who enjoy strong sensations seek this discharge of adrenaline. Is that why Westerners are increasingly fond of hot food, whether Asian or Mexican? In the southern United States, chili peppers are almost a staple. Eating spicy food is a sign of strength, endurance, and virility. Hot and spicy: it's easy to see a parallel between sexual and gustatory pleasures. In an episode of *The Simpsons*, Homer Simpson coats his throat with hot wax so he can swallow a hot pepper from the Guatemalan jungle.

A scale measuring the amount of capsaicin in peppers was invented by the American chemist Wilbur Scoville in 1912. A mere 1 percent of this substance equals 150,000 units of heat, which is already above the count in a spicy Thai dish. The "red savina" in California is second only to the chili. In September 2000, a military laboratory in northeastern India announced it had found the hottest pepper in the world. The *bhut jolokia* was tested at the Chili Pepper Institute in New Mexico, home to societies and journals devoted to the chili pepper: to everyone's surprise, the *bhut jolokia* reached more than one million units of heat. It is eaten by the Nagas, a tribe that named the pepper. It is currently being used for lachrymogenic gases as well as to keep elephants out of villages. We know that sprays used for self-defense often contain concentrated capsaicin. But its heating qualities are also used in medicines to treat sprains, twisted joints, or athletes' muscles. Very rich in vitamins, the chili pepper was also the source of Albert Szent-Györgyi's work on vitamin C in 1928; this biochemist, an American born in Budapest, succeeded in isolating and crystallizing the vitamin, for which he was awarded the Nobel Prize.

The genetic map of the chili pepper is progressing. Today we know it as a mine of antioxidants; centuries ago the Incas knew it as a "source of joy and beauty." Its spread throughout the world can hardly surprise us.

✣ ✣ ✣ ✣ ✣ ✣ ✣ ✣ ✣ ✣ ✣ ✣ ✣ ✣ ✣ ✣ ✣ ✣ ✣ ✣ ✣

### LE VERGER

Dans le jardin, sucré d'oeillets et d'aromates,
Lorsque l'aube a mouillé le serpolet touffu,
Et que les lourds frelons, suspendus aux tomates,
Chancellent, de rosée et de sève pourvus,

Je viendrai, sous l'azur et la brume flottante,
Ivre du temps vivace et du jour retrouvé,
Mon coeur se dressera comme le coq qui chante
Insatiablement vers le soleil levé.

L'air chaud sera laiteux sur toute la verdure,
Sur l'effort généreux et prudent des semis,
Sur la salade vive et le buis des bordures,
Sur la cosse qui gonfle et qui s'ouvre à demi;

La terre labourée où mûrissent les graines
Ondulera, joyeuse et douce, à petits flots,
Heureuse de sentir dans sa chair souterraine
Le destin de la vigne et du froment enclos.

Des brugnons roussiront sur leurs feuilles, collées
Au mur où le soleil s'écrase chaudement;
La lumière emplira les étroites allées
Sur qui l'ombre des fleurs est comme un vêtement.

Un goût d'éclosion et de choses juteuses
Montera de la courge humide et du melon,
Midi fera flamber l'herbe silencieuse,
Le jour sera tranquille, inépuisable et long.

Et la maison, avec sa toiture d'ardoises,
Laissant sa porte sombre et ses volets ouverts,
Respirera l'odeur des coings et des framboises
Éparse lourdement autour des buissons verts;

Mon coeur, indifférent et doux, aura la pente
Du feuillage flexible et plat des haricots
Sur qui l'eau de la nuit se dépose et serpente
Et coule sans troubler son rêve et son repos.

Je serai libre enfin de crainte et d'amertume,
Lasse comme un jardin sur lequel il a plu,
Calme comme l'étang qui luit dans l'aube et fume,
Je ne souffrirai plus, je ne penserai plus,

Je ne saurai plus rien des choses de ce monde,
Des peines de ma vie et de ma nation,
J'écouterai chanter dans mon âme profonde
L'harmonieuse paix des germinations.

Je n'aurai pas d'orgueil, et je serai pareille,
Dans ma candeur nouvelle et ma simplicité,
A mon frère le pampre et ma soeur la groseille
Qui sont la jouissance aimable de l'été,

Je serai si sensible et si jointe à la terre
Que je pourrai penser avoir connu la mort,
Et me mêler, vivante, au reposant mystère
Qui nourrit et fleurit les plantes par les corps.

Et ce sera très bon et très juste de croire
Que mes yeux ondoyants sont à ce lin pareils,
Et que mon coeur, ardent et lourd, est cette poire
Qui mûrit doucement sa pelure au soleil ...

\* Anna de Noailles, from *Le Coeur innombrable* (1901)

## THE ORCHARD

In the garden, sweetened with carnations and herbs,
When dawn has dampened the dense thyme
And heavy drones, hanging from the tomatoes,
Stagger, sated with dew and sap,

I will come, under the blue sky and floating mist,
Intoxicated by the newfound life and day,
My heart will rise up like the crowing cock
Insatiably toward the rising sun.

The warming air flows like milk over the green,
Over the generous and careful effort of the seedlings,
Over the lettuce and the bordering boxwood,
Over on the pods that swell and open halfway;

The plowed field where seeds are ripening
Will roll along like the sea, joyful and sweet, in little waves,
Happy to feel in its underground flesh
The destiny of the vine and the wheat within.

Nectarines will turn red on the leaves, clinging
To the wall where the sun is warmly striking;
The light will fill the narrow lanes
On which the flowers cast their shadow like a garment.

A sense of hatching and of juicy things
Will climb from the wet squash and melon,
Noon will set ablaze the silent grass,
The day will be quiet, inexhaustible, and long.

And the house, with its slate roof,
Leaving its dark door and shutters open
Will breathe the scent of quince and raspberries
Scattered heavily around the green bushes;

My heart, indifferent and soft, will bend
Toward the flexible, flat leaves of the beans
On which the water of the night rests and trickles down
And flows without disturbing its dream or its rest.

I will finally be free of fear and bitterness,
At peace like a garden that has been rained upon,
Calm like the gleaming, misty pond,
I will no longer suffer, I will no longer think,

I will know nothing more of the things of this world,
The sorrows of my life or my nation,
I will listen to the singing in my deep soul
The harmonious peace of germination.

I will have no pride, I will resemble,
In my new innocence and simplicity,
My brother the vine and my sister the currant
Who are the sweet enjoyment of summer

I will be so sensitive and so attached to the land
That I will be able to think I have known death,
And to blend, alive, with the restful mystery
That feeds and nourishes plants, giving them body.

And it will be very good and right to believe
That my glistening eyes are like this flax,
And that my heart, heavy and on fire, is that pear
Gently ripening its skin in the sun . . .

\* [My translation.—TRANS.]

✢ ✢ ✢ ✢ ✢ ✢ ✢ ✢ ✢ ✢ ✢ ✢ ✢ ✢ ✢ ✢ ✢ ✢ ✢ ✢ ✢ ✢ ✢ ✢

## ACKNOWLEDGMENTS

I wish to thank Michel Onfray and the entire team at the Université populaire d'Argentan.

I thank all those who have followed me in these "histories of taste."

I thank Christiane Alexis, Christiane Dorléans, and Fabrice Egler.

I thank my publisher and everyone at Grasset who contributed to the publication of this work.

# A Biographer of Vegetables

MICHEL ONFRAY

✤ ✤ ✤ ✤ ✤ ✤ ✤ ✤ ✤ ✤ ✤ ✤ ✤ ✤ ✤ ✤ ✤ ✤ ✤ ✤ ✤ ✤ ✤ ✤

## Biographies of a Biographer

As I was thinking of friends to invite to the Fourieristic meal of what was to become the Université populaire du goût (UPG) in Argentan,* I decided to ask Évelyne Bloch-Dano. I asked her as a friend I knew I could count on, but also as a biographer. First as a friend, for I remembered our conversations on Sartre and Beauvoir, our shared passion for the socialist writer Flora Tristan, the homes of writers as fragments of their work, then Stendhal's development; respect for truth in film, Proust's style, Colette's style, and many other interests we shared and on which our friendship was based.

I also called upon the biographer. We had had many conversations about biography, writing about oneself, autobiography, the truth of a life, the construction of a self, the Sartrean project of existential psychoanalysis, Montaigne's genius, the blind spot of an existence, things to be said or not to be said, things to show, once-in-a-lifetime moments, the complex relationship between the biographer and her subjects, the autobiography that is always implied in choosing to write one biography rather than another, and so on.

---

*"Fourieristic": from the noun Fourierism, the social system proposed by François-Marie-Charles Fourier (1772–1837), under which society was to be organized into phalanxes or associations, each large enough for all industrial and social requirements. Fourier declared that concern and cooperation were the secrets of social success. He believed that a society that cooperated would see an immense improvement in its productivity levels. Université populaire du goût: "People's University of Taste."

And Évelyne Bloch-Dano is a biographer. That is self-evident in her specific biographies—of Zola's wife, Proust's mother, and Flora Tristan—but it is also the case in her work on writers' houses, or even when she tries her hand at fiction, in which, once again, she becomes the biographer of her father, her mother, and of course herself, but equally of Romy Schneider as her double, her shadow, as imitations (in the Epicurean sense) of her parents. How could she ever escape her destiny as a biographer? How could she avoid encountering her own life while looking at those of others?

*The Voice of Vegetables*

Having called upon my friend, and hoping to rally other friends around the principle of group enjoyment so dear to Fourier, I asked the biographer to present to the audience at the UPG nothing less than a biography of vegetables. For the UPG was planning to initiate "Gardens in the City," a project of social reintegration that would employ a dozen or so victims of socioeconomic exclusion who had hit rock bottom, in order to help them regain their dignity.

The vegetable garden can serve as a springboard for rediscovering lost dignity: the idea isn't to train experienced horticulturalists, expert gardeners, or productive vegetable merchants, but to enable women and men to rediscover their self-esteem. The friend who would oversee this garden was himself a victim of the capitalist marketplace: after working for fifty years in the same factory, he was laid off when his company moved to China, and reintegrated into . . . reintegration.

When he told me that bushel baskets of vegetables offered to the food bank have no takers because potential recipients either cannot or will not cook them, I decided to create the UPG. As with most things, I couldn't have done it alone. Cooking classes were given by master chefs who volunteered their time, led by Marc de Champérard, who, using persuasion and rhetoric, inspired the upper crust of French chefs to create variations on the seven vegetables chosen for as many sessions.

Évelyne Bloch-Dano responded immediately. She was generous, available, ready to officiate in front of more than five hundred people, in the steam arising from the pots in which the chefs prepared their dishes, and to recount, as a biographer, the adventures of those vegetables, which had suddenly become characters in a novel, heroes in a film, players on the stage of planetary geography, cosmopolitan actors, familiar faces. As La Fontaine did with animals, she gave voice to a parsnip or to a tomato, and gave the floor to the vegetables being honored in the banquet hall of the sous-préfecture.

## Through the Gates of the Vegetable Garden

How did she do this? By showing, as only a biographer can, how one becomes what one is—when one is a pea, a bean, a Jerusalem artichoke. In other words, by showing that a vegetable possesses a symbolic aura that is greater than its caloric or market value. Or by revealing its poetry as well as its genetics. By telling the odyssey, the destiny, of a vegetable, each one unique, each one different, each a variation of the same: She presented *the* tomato, and the *specific* tomatoes that were being cooked that day.

To tell that adventure is to enter into universal history through the gate to the kitchen garden. First you greet the gardener, and then you encounter Hegel among the peas. For the most modest of vegetables contains a universal adventure. (It should be pointed out that the very term "modest vegetable" is a redundancy. The truffle, for example, immodest through no fault of its own, is never classified among vegetables. Nor among mushrooms. It escapes taxonomy.)

When you eat a vegetable, then, what you are ingesting is the history of the world. It is the biographer's task to tell us when and how. So Évelyne Bloch-Dano invoked a plethora of disciplines: literature, art history, music, poetry, film, history, prehistory, geography, geology, geomorphology, climatology, genetics, horticulture, gardening theory . . .

The sessions at the UPG showed first that frontiers, nations, and countries derive from the madness of humanity; a vegetable recognizes only the soil and the climate that allow it to grow. Second, however, they showed that the history of nature becomes the history of what people have done with it. So it had to be told how these plants, uprooted and moved, experienced another destiny, which corresponded to the writing of another page in their great novel. The biographer of a vegetable puts herself at the service of this epic of which she becomes the memorialist.

## Affective Calories

Every vegetable contains two histories: the large, the one that enabled Hegel to write *Reason in History*; and the small, the one contained in the memories of each one of us. The large history concerns the conquistadors, the spice route, caravels and the opening of maritime passages, the commerce of empires, the cycle of exports and imports, the arrival of a vegetable into a country, economy, diplomacy, continental politics, and so forth. The small history is always that of fathers and mothers, grandfathers and grandmothers.

Scratch the gold on the crown of a world-famous chef, the one with the most stars, the most eminent, the most recognized, whether he is friendly or surly, modest or arrogant, discreet or pretentious, and you will find the same gems as those in the crown of the cook in a small country restaurant, seldom seen, barely known. And those same gems are found among impassioned cooks, male and female, and in the pantry in which Proust's madeleines are stored.

What makes a person want to cook? It is the desire to rediscover the tastes and smells of childhood, of those times when we were blissfully unaware of our mortality, and which transformed each second into a springtime lunch. Deep within most of us, near the surface for others, are the seeds of smells, diamonds, perfumes, gems, tastes, constituting so many gustatory treasures. The taste of a blackberry plucked in the woods, the crunchy texture of a nut gathered on a footpath, the tang of an apple stolen from the neighbor's orchard, the taste of the juice in a blade of grass or a buttercup in your mouth.

What made each of Evelyn Bloch-Dano's sessions so interesting was her demonstration that we can not only stuff ourselves like any pig, but can also *taste*—that while we know we are consuming dietary calories, we are also ingesting affective calories that can be rediscovered—if we know they exist, if we want to encounter them, and if we realize that we must go find them. This is the meaning of the UPG: to teach in the form of an appetizer that it is possible for everyone to rediscover the pleasure of a complete childhood by *tasting* what we eat in order to bring out the poetry that is always hidden within.

This is exactly what the social theorist Charles Fourier taught with his gastrosophy in the early nineteenth century: to make food (and opera, but that's another story) a learning experience, a path into another world, in which pleasure is not a sin but the cement of a new community. The UPG offers a type of micro-gastrosophic republic in which Évelyne Bloch-Dano holds a cardinal position. Fourier would have been delighted that a woman should play this role: he believed, indeed, that true revolution would be brought about by women active in his utopian "phalanxes." Let's forget Marx and read or reread Fourier—and Évelyne Bloch-Dano.

# Bibliography

✛ ✛ ✛ ✛ ✛ ✛ ✛ ✛ ✛ ✛ ✛ ✛ ✛ ✛ ✛ ✛ ✛ ✛ ✛ ✛ ✛ ✛ ✛ ✛ ✛

## Reference Works

Avry, Marie-Pierre, and François Galloin. *Légumes d'hier et d'aujourd'hui*. Belin, 2007.

*Jardins savoureux en pays d'Auge*. Association Monviette Nature, 2001.

Lemoine, Elizabeth. *Les Légumes d'hier et d'aujourd'hui*. Éditions Molière, 2003.

Pelt, Jean-Marie. *Des légumes*. Fayard, 1993.

———. *Ces plantes que l'on mange*. Éditions du Chêne, 2006.

Pitrat, Michel, and Claude Foury. *Histoires de légumes des origines à l'orée du XXI$^e$ siècle*. Éditions INRA, 2003.

Vivier, Michel. *Savoirs et secrets des jardiniers normands*. Éditions Charles Corlet, 2007.

## Basic Texts

Apicius. *In re coquinaria*. Edited and translated by Christopher Grocock and Sally Grainger as *Apicius: A Critical Edition with an Introduction and an English Translation of the Latin Recipe text Apicius*. Prospect, 2006.

Bonnefons, Nicolas de. *Délices de la campagne*. Edited by Pierre de Gagnaire and Hervé This. In *Alchimistes aux fourneaux*. Flammarion, 2007.

———. *Le Jardinier français* (1651). Edited by François-Xavier Bogard. Ramsay, 2001.

Brillat-Savarin, Jean Anthelme. *La Physiologie du goût* (1825). Translated by M. F. K. Fisher as *The Physiologie of Taste*. Knopf, 1971.

Dumas, Alexandre. *Grand Dictionnaire de cuisine*. Lemerre, 1873.

Grimod de La Reynière, Alexandre-Balthazar-Laurent (1758–1837). *Écrits gastronomiques*. 10/18, 1978.

*Le Mesnagier de Paris* (circa 1393).

Platina. *De honesta voluptate et valetudine* (ca. 1470). Translated by Mary Ella Milham as *On Right Pleasure and Good Health*. Medieval and Renaissance Texts and Studies, 1998.

Pliny the Elder. *The Historie of the World; commonly called The Naturall Historie of C. Plinius Secundus*. Translated by Philemon Holland. 1635.

Serres, Olivier de. *Théâtre d'agriculture et mesnage des champs*. 1600.

Taillevent. *Le Viandier de Taillevent: 14th Century Cookery, Based on the Vatican Library manuscript* (1370). Translated by James Prescott. 2nd ed. Alfarhaugr Publication Society, 1989.

## History of Food

Barrau, Jacques. *Les Hommes et leurs aliments*. Messidor / Temps Actuels, 1983.

Flandrin, Jean-Louis, and Massimo Montanari, eds. *Histoire de l'alimentation*. Fayard, 1996.

Garine, Jean-Louis. "Les modes alimentaires, histoire de l'alimentation et des manières de table." In *Histoire des moeurs*, Encyclopédie de la Pléiade. Gallimard, 1990.

Meiller, Daniel, and Paul Vannie. *Le Grand Livre des fruits et légumes*. Éditions La Manufacture, 1991.

OCHA: Website on food culture and behaviors. www.lemangeur-ocha.com.

Toussaint-Samat, Maguelonne, *A History of Food*. Translated by Anthea Bell. Blackwell Reference, 1992.

## Monographs

Aron, Jean-Paul, *Le Mangeur du XIXᵉ siècle*. Robert Laffont, 1973.

Baraton, Alain. *Le Jardinier de Versailles*. Grasset, 2006.

Boudan, Christian. *Géopolitique du goût: La Guerre culinaire*. Presses universitaires de France, 2004.

Bruegel, Martin, with Bruno Laurioux. *Histoire et identités alimentaires en Europe*. Hachette Littérature, 2002.

Capatti, Alberto, and Massimo Montanari. *Italian Cuisine: A Cultural History*. Translated by Aine O'Healy. Columbia University Press, 2003.

Ferrières, Madeleine. *Histoire des peurs alimentaires: Du Moyen Âge à l'aube du XXᵉ siècle*. Seuil, 2002.

Gillet, Philippe. *Le Goût et les Mots: Littérature et gastronomie (XIVᵉ-XXᵉ siècle)*. Petite Bibliothèque Payot, 1993.

Laurioux, Bruno. *Manger au Moyen Âge*. Pluriel, Hachette Littérature, 2002.

Poulain, Jean-Pierre, and Edmond Neirinck. *Histoire de la cuisine et des cuisiniers: Techniques culinaires et pratiques de table, en France, du Moyen Âge à nos jours*. LT Jacques Lanore, 2004.

Revel, Jean-François. *Un festin en paroles*. Pluriel, Pauvert, 1979.

Rowley, Anthony. *Une histoire mondiale de la table: Stratégies de bouche*. Odile Jacob, 2006.

### Imagery

Impelluso, Lucia. *Gardens in Art*. Translated by Stephen Sartarelli. J. Paul Getty Museum, 2007.

Malaguzzi, Silvia. *Food and Feasting in Art*. Translated by Brian Phillips. J. Paul Getty Museum, 2008.

### Terminology

Duneton, Claude. *Le Bouquet des expressions imagées*. Seuil, 1990.

Guillemard, Colette. *Les Mots d'origine gourmande*. Belin, 1988.

Onfrey, Michel. *La Raison gourmande*. Grasset, 1995.

———. *Le Ventre des philosophes*. Grasset, 1989.

Rey, Alain. *Dictionnaire culturel en langue française*. Le Robert, 2005.

### Supplementary Sources

GOOD TO EAT, GOOD TO THINK ABOUT

Aragon, Louis. *Traité du style*. Gallimard, 1928. Translated by Alyson Waters as *Treatise on Style*. University of Nebraska Press, 1991.

Lévi-Strauss, Claude. *Du miel aux cendres* (1967). Translated by John and Doreen Weightman as *From Honey to Ashes*. Cape, 1973.

———. *L'origine des manières de table* (1968). Translated by John and Doreen Weightman as *The Origin of Table Manners*. Harper & Row, 1978; repr., University of Chicago Press, 1990.

———. "Le triangle culinaire." *L'Arc*, no. 26 (1965).

Ronsard, Pierre de. *Poésies*. 1569.

Zola, Émile. *L'Assommoir*. Translated by Margaret Mauldon. Oxford University Press, 1998.

———. *Le Ventre de Paris* (1873). Translated by Brian Nelson as *The Belly of Paris*. Oxford University Press, 2008.

A MATTER OF TASTE

Boileau, Nicolas. *Satires et épîtres*, satire III. 1665.

Cornaro, Nicolas. *Discorsi della vita sobria* (1558). Translated as *The Art of Living Long*. W. F. Butler, 1905.

Diderot, Denis. *Lettre sur les aveugles à l'usage de ceux qui voient* (1749). Translated as *An Essay on Blindness, in a letter to a person of distinction*. 1780.

Erasmus of Rotterdam. *De civitate morum puerilium* (1530). Translated by Eleanor Merchant as *A Handbook on Good Manners for Children*. Preface Publishing, 2008.

Flandrin, Jean-Louis. "La Distinction par le goût." In *Histoire de la vie privée 3: De la Renaissance aux Lumières*, edited by Philippe Ariès and Georges Duby. Points, Seuil, 1999.

Montaigne, Michel de. *Essais*, I, 51, *De la vanité des paroles*. 1580.

Voltaire. *Dictionnaire philosophique*. 1765.

THE CARDOON AND THE ARTICHOKE

Bresson, Aïté. *L'Artichaut et le Cardon*. Chroniques du potager (Actes Sud), 1999.

Freud, Sigmund. *The Interpretation of Dreams*. Translated by Joyce Frick. Oxford University Press, 1999.

Gibault, Georges. "Le cardon et l'artichaut." In *Histoire des légumes*. Paris, 1907.

Goethe, Johann Wolfgang von. *Italian Journey: 1786–1788*. Translated by W. H. Auden and Elizabeth Mayer. Penguin Classics, 1992.

Morren, Charles. *Palmes et couronnes de l'horticulture de Belgique*. 1851.

Proust, Marcel. *Du côté de chez Swann* (1913). Translated by C. K. Scott Moncrieff as *Swann's Way* (1928). Dover Publications, 2002.

THE JERUSALEM ARTICHOKE

Hennig, Jean-Luc. *Le Topinambour et autres merveilles*. Zulma, 2000.

THE CABBAGE

Delteil, Joseph. *Les Cinq Sens*. Denoël, 1983.
———. *La Cuisine paléolithique*. Les éditions de Paris, 2007.
Jonasz, Allen S. *La Famille*. Paroles et musique Michel Jonasz. Éditions Marouani, Wea Music, 1977.
Gainsbourg, Allen S. *L'Homme à tête de chou*. Paroles et musique Serge Gainsbourg. Philips, 1976.
Le Roy, Eugène. *Jacquou le Croquant*. Le Livre de poche, 1997.
Thorez, Jean-Paul. *Les Choux*. Chroniques du potager (Actes Sud), 2002.
Vivier, Michel. *Jardins ruraux en Basse-Normandie*. Centre régional de culture ethnologique et technique de Basse-Normandie, Caen, 1998.
Weiss, Allen S. *Autobiographie dans un chou farci*. Mercure de France, 2006.

THE PARSNIP

Beckett, Samuel. *Premier amour* (1945). Translated by the author as *First Love*. Calder & Boyars, 1973.
Céline, Louis Ferdinand. *Mort à crédit*. Gallimard, 1952.

THE CARROT

Goust, Jérôme. *La Carotte et le Panais*. Chroniques du potager (Actes Sud), 2001.
Proust, Marcel. Lettre à Céline Cottin. In *Correspondance*, vol. 9. Plon, 1982.

THE PEA

Andersen, Hans Christian. "The Princess and the Pea." 1835.
Delerm, Philippe. *La Première Gorgée de bière*. L'Arpenteur, 1997. Translated as *We Could Almost Eat Outside: An Appreciation of Life's Small Pleasures*. Picador, 1999.
Jacobsohn, Antoine, and Dominique Michel. *Le Petit Pois*. Chroniques du potager (Actes Sud), 2001.
Platina. *Pois de lard*. Cited by Philippe Gillet in *Le Goût et les Mots* (see under Monographs, above).

THE TOMATO

Daneyrolles, Jean-Luc. *La Tomate*. Chroniques du potager (Actes Sud), 1999.
Delteil, Joseph. *Choléra*. In *Oeuvres complètes*. Grasset, 1961.
———. *La Cuisine paléolithique*. Les éditions de Paris, 2007.
Prior, Lily. *La Cucina: A Novel of Rapture*. Harper Perennial, 2001.

THE BEAN

Allende, Isabel. *Aphrodita: A Memoir of the Senses*. Translated from the Spanish by Margaret Sayers Peden. HarperFlamingo, 1998.
Goust, Jérôme. *Le Haricot*. Chroniques du potager (Actes Sud), 1998.
Revel, Jean-François. *Un festin en paroles*. Pluriel, 1982.
Steinbeck, John. *Tortilla Flat*. 1935.
Thoreau, Henry David. *Walden*. 1854.

THE PUMPKIN

Bresson, Aïté. *Le Potiron*. Chroniques du potager (Actes Sud), 1998.
Coffe, Jean-Pierre. *Le Bon Vivre*. Le pré aux clercs, 1989.
Pahud, Yvonne, Marinette Tardy, and Martine Meldem. *Courge, citrouille et potiron: Saveurs gourmandes*. Cabédita, 2006.
Polet, Jean-Claude, ed. *Patrimoine littéraire européen 5—Anthologie en langue française*. De Boeck Université, 1992.

THE CHILI PEPPER

Daneyrolles, Jean-Luc. *Le Piment et le Poivron*. Chroniques du potager (Actes Sud), 2000.
Robinson, Simon. "Global Warming." *Time Magazine*, June 25–July 2, 2007.